Listening to Longwave

The World Below 500 Kilohertz

First Edition   By Kevin Carey

Universal Radio Research
6830 Americana Parkway
Reynoldsburg, Ohio 43068
United States of America

This book is largely adapted from
The World Below 500 Kilohertz
By
L. Peter Carron Jr.

Cover:
An extremely simplified diagram of the radio spectrum

FIRST EDITION

FIRST PRINTING

Copyright ©2007
Fred Osterman

Universal Radio Research
6830 Americana Parkway
Reynoldsburg, Ohio 43068

Reproduction or publication of the contents in any manner without the written permission of the publisher is prohibited. No liability is assumed with respect to the use of information herein.

Printed in the United States of America

ISBN 1-882123-39-5

CONTENTS

	Preface	iv
1	Overview of the Longwave Spectrum	1
2	What Transmits Where	3
3	Propagation Conditions	8
4	Common Users of Longwave	9
5	Special Users	32
6	Lowfers	44
7	Natural Longwave Signals	50
8	Unusual Uses for the Longwaves	57
9	Receivers & Converters for Longwave	62
10	Longwave Lowfer Transmitters	72
11	Longwave Antennas	74
12	Listening Tips	80
13	References	89

PREFACE

The little-known, seldom-listened-to longwave portion of the radio spectrum below the AM radio band contains a fascinating collection of signals. Nowhere else can you find an area of the spectrum that houses such a diverse and unusual group of emissions – ranging from aircraft beacons to communications with underwater submarines to signals emitted by the ionized trails of missile launches. In some parts of the world, there are even broadcast stations. In addition, there are a host of mysterious signals emitted by the Earth itself, known by intriguing names like sferics, tweeks, whistlers, and dawn chorus.

There are technical books that explain how to construct electronic gear for receiving and transmitting in this part of the radio spectrum. This book is not technical. It tells the intriguing story of longwave and explores the best way to hear these transmissions.

Much of the early experimentation in radio took place on the longwaves. This photo shows an early LF transmitter used by Hiram P. Maxim, founder of the American Radio Relay League. It operated in the Low Frequency part of the longwave spectrum (see next page). "Old Betsy," as it is known to staffers, is on display at the League's headquarters in Newington, Connecticut.

Chapter 1
Overview of the Longwave Spectrum

The radio spectrum is divided into nine different areas, each spanning a certain frequency range. These divisions are as follows:

ELF	Extremely Low Frequencies	0 kHz - 3 kHz
VLF	Very Low Frequencies	3 kHz - 30 kHz
LF	Low Frequencies	30 kHz - 300 kHz
MF	Medium Frequencies	300 kHz - 3 MHz
HF	High Frequencies	3 MHz - 30 MHz
VHF	Very High Frequencies	30 MHz - 300 MHz
UHF	Ultra High Frequencies	300 MHz - 3 GHz
SHF	Super High Frequencies	3 GHz - 30 GHz
EHF	Extremely High Frequencies	30 GHz - 300 GHz

The abbreviations used for frequency definition are:

1 Hz	= 1 Hertz	= 1 cycle per second	
1 kHz	= 1 kilohertz	= 1,000 Hertz	
1 MHz	= 1 Megahertz	= 1,000 kilohertz	= 1,000,000 Hertz
1 GHz	= 1 Gigahertz	= 1,000 Megahertz	= 1,000,000,000 Hertz

This book is concerned with the extent of frequencies between 0 kHz and 500 kHz which encompasses the Extremely Low Frequencies, Very Low Frequencies, Low Frequencies and portions of the Medium Frequencies. For ease of presentation, however, this portion of the spectrum will be referred to as the "Low Frequencies" or the "longwaves."

To whet your interest in this unusual area, listed here is a sampling of what can be heard with simple monitoring equipment:

- Navigation beacons
- Low frequency experimenters
- Standard frequency and time stations
- Morse code
- Radioteletype
- Marine transmissions
- Foreign broadcasts
- Global navigation signals
- VLF/LF communications with submarines
- Weather stations
- Military transmissions
- Natural radio signals such as sferics, whistlers, tweeks and dawn chorus

Chapter 2
What Transmits Where

To explain what services are permitted to transmit where, a simplified chart of the longwave spectrum is shown. The allocation of frequencies is quite complex. This task is carried out on an international scale by the International Telecommunications Union (ITU) and within the United States by the Federal Communications Commission (FCC). There are many territorial exceptions, notations, restrictions, and other stipulations. An attempt to show all variations would defeat the purpose of this book, which is to provide an easy-to-understand reference document.

What follows is a summary of activity on the longwave band from 0 to 535 kHz, accompanied by a chart showing the prominent users of the band. Every reasonable effort has been made to ensure the accuracy of this information at the time of writing, but radio frequency usage is constantly in a state of flux. Stations come and go and frequencies may change, often without notice to the general public.

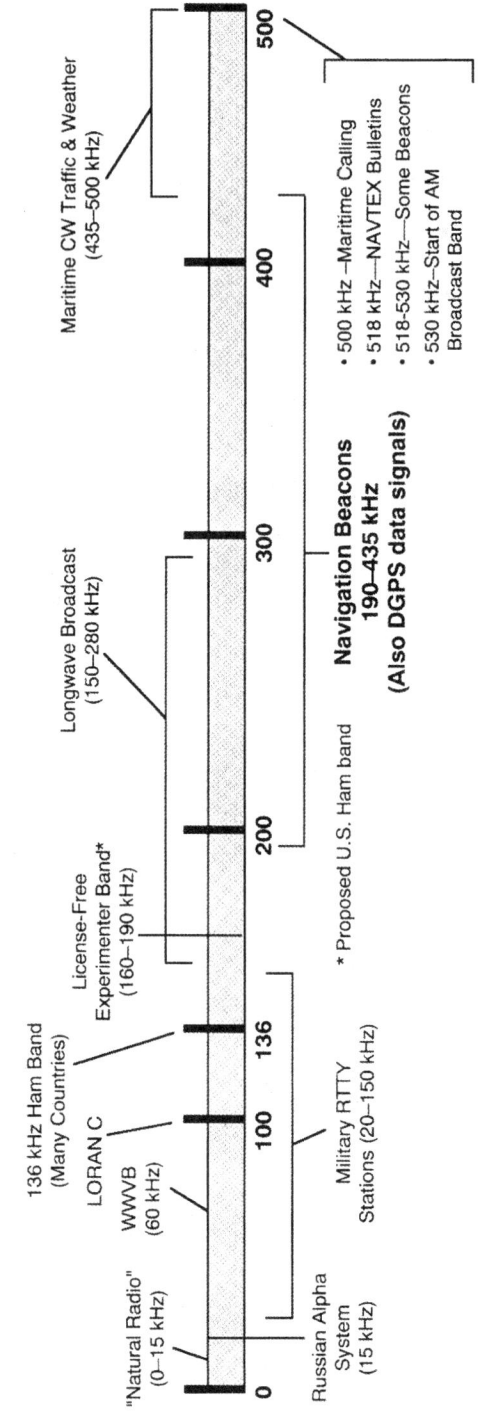

0 - 10 kHz

For the most part, this is the territory of "natural radio." Listeners with specialized, yet uncomplicated equipment can hear signals in this range that are generated by the Earth itself. These signals include strange-sounding sferics, whistlers, tweeks, and dawn chorus. They are believed to be caused by the interaction of three factors: lightning, the Earth's magnetic field, and sunspot activity. Until recently, there was one prominent man-made signal at the bottom of this range—the U.S. Navy ELF system operating at 76 Hz (that's less than 1 kHz!). This system consisted of two stations—one at Clam Lake, Wisconsin, and another at Republic, Michigan. This system was used for transmission to submerged submarines. The military recently announced the decommissioning of this system after more than 30 years of operation.

10 - 15 kHz

This range is allocated to radio navigation services. For many years the beeps of the now-defunct OMEGA system could be heard 24 hours a day, providing high-accuracy navigation signals to mariners and aviators worldwide. On September 30, 1997, the system was shut down by the U.S. Coast Guard in favor of the satellite-based Global Positioning System (GPS), which offers better accuracy and availability. As of this writing, a Russian radiolocation system called ALPHA remains active near 15 kHz. ALPHA signals are frequently heard by North American listeners.

15 - 150 kHz

A number of military FSK (frequency shift keying) stations operate in this range. Most use high power and massive antenna systems to provide global or continental coverage—even during daylight hours. As you might have guessed, their signals are heavily encrypted. Besides FSK, you may hear CW traffic on these frequencies, although this has become increasingly rare.

At 60 kHz is time station WWVB, Fort Collins, Colorado. This is a sister station of well known HF time stations WWV and WWVH. Another point of interest is 136 kHz. Many countries allow ham or experimental operation on this frequency under strict power and antenna limitations.

150 - 190 kHz
This frequency range is significant for two reasons. First, it marks the start of the LF broadcast band (150-280 kHz) for ITU Region 1 which encompasses Europe, Africa, the Middle East and Russia. Second, it is home to the license-free experimenter's band authorized in the United States and Canada (160 –190 kHz).

The experimenter's band is home to a hardy group of operators who use mostly home-built equipment to communicate with other enthusiasts. Current regulations in the U.S. and Canada limit transmitters to a maximum power output of 1 watt and antennas to a maximum length of 15 meters, or about 50 feet (including the feedline). At this writing, a proposal is being considered by the FCC that would permit licensed ham operations in this band and at 136 kHz. Approval for at least one of these frequency ranges is considered likely.

Starting at 150 kHz, and extending to 280 kHz is the longwave broadcast band used in Europe, the Middle East, Africa, and Asia. LW broadcasters generally run very high power and are frequently heard by listeners in North America when conditions are favorable. Reception is most likely when a path of darkness exists between the transmitting site and the receiving location.

190 - 435 kHz
For most longwave listeners, this frequency range is considered a mainstay of their DXing activities. The band is sprinkled with unmanned, low-power (50 watts or less) beacon stations transmitting their IDs over and over again in slow Morse Code. Intended mainly for homing by pilots, these stations also serve as an adjunct to more sophisticated systems such as VOR and radar. Because most beacons operate 24 hours a day from a known location, they make ideal DX targets.

435 - 530 kHz

Until the mid-1980s, this part of the band was busy with ship-to-shore CW traffic and marine weather transmissions. With the advent of VHF and satellite communications, however, activity has been sparse in recent times. There is still some activity to be heard, but the majority of transmissions are one-way broadcasts, such as weather forecasts and iceberg warnings. In time, even these stations will likely fade away.

At exactly 500 kHz is the former International Distress and Calling frequency. Although many nations have stopped official monitoring, the channel is still used occasionally for establishing contact between ship and shore stations.

The FCC's Office of Engineering and Technology granted a Part 5 experimental license WD2XSH to the ARRL on September 13, 2006 on behalf of a group of radio amateurs interested in investigating spectrum in the vicinity of 500 kHz. The two-year authorization permits experimentation and research between 505 and 510 kHz (600 meters) using narrowband modes at power levels of up to 20 Watts effective radiated power (ERP). ARRL Member Fritz Raab, W1FR, of Vermont, is the project manager for *The 500 KC Experimental Group for Amateur Radio*. See www.500kc.com for further information on the 600 meter experimental group.

A major exception to the decline in maritime activity can be found at 518 kHz. This is the home of NAVTEX—a worldwide marine teleprinting service that broadcasts safety and navigation information to ocean-going vessels. NAVTEX signals can be decoded and read with inexpensive digital receiving equipment.

A few navigation beacons operate between 518 and 530 kHz, the bottom end of the AM broadcast band.

Chapter 3
Propagation Conditions

Longwave propagation conditions differ from those in any other portion of the spectrum. The unique characteristics make the band useful for certain types of transmissions.

Primarily, radio waves here tend to be very stable, with less fluctuation in signal strength than those found elsewhere. Near the upper end of the longwave spectrum, signals tend to be less stable, acting much like AM broadcast stations. Most noticeably, stations are much stronger at night than during the day.

Lower down, however, radio waves tend to propagate by hugging the Earth. This results in a more stable signal and one that is less absorbed by both the layers of the ionosphere and the Earth itself. As a matter of fact, go low enough in frequency and waves will actually penetrate the Earth – or water! This characteristic will be discussed at greater length in upcoming chapters.

The Very Low Frequencies have drawbacks, however. Most prominent of these is noise. Lightning and other natural phenomena generate signals at the low end of the spectrum. Summer storms wreak havoc on listening. And man-made noise can be just as bad, especially in urban environments. Electric machines, light-dimmer controls, switches, power lines, computers, televisions and a multitude of other modern conveniences all add their part to the chorus of static that can, at times, destroy longwave reception.

Later in this book we'll discuss ways of identifying longwave noise and provide some tips for reducing it to an acceptable level.

Chapter 4
Common Users of the Longwaves

Chapter 1 gave an overall view of the longwave spectrum and listed some of its users. This chapter and the next three chapters are devoted to an in-depth look at occupants of this "basement band."

BEACONS

Beacons are electronic lighthouses. They are used to pinpoint such things as airports, offshore drilling platforms and other points of reference for air and sea navigation. Most beacons operate 24 hours a day, transmitting an identifying signal again and again in slow Morse code, giving listeners a target to home in on. The majority use a simple identifier of one to three characters. A few different beacons use the same identifier, usually when the transmitter locations are separated far apart geographically or the frequencies used are so different that it is easy to distinguish one from another.

Although configurations vary, many beacons are housed in small, almost square buildings, and are frequently surrounded by a fence to ward off curiosity seekers. The outside antenna can be a vertical top-hat arrangement, or possibly a flat-top "T" antenna. The latter is most common at older beacon sites. Chances are you have seen one of these situations near an airport and perhaps didn't realize what it was.

```
                    MORSE CODE
        A  •—           N  —•           1  •————
        B  —•••         O  ———          2  ••———
        C  —•—•         P  •——•         3  •••——
        D  —••          Q  ——•—         4  ••••—
        E  •            R  •—•          5  •••••
        F  ••—•         S  •••          6  —••••
        G  ——•          T  —            7  ——•••
        H  ••••         U  ••—          8  ———••
        I  ••           V  •••—         9  ————•
        J  •———         W  •——          0  —————
        K  —•—          X  —••—
        L  •—••         Y  —•——
        M  ——           Z  ——••
```

Beacons transmit in slow Morse code.

Typical beacon station with "Tophat" antenna.
This is EVB 417 kHz in New Smyrna Beach, Florida.

A sign, such as this typically greets visitors at NDB (Non-Directional Beacon) sites.

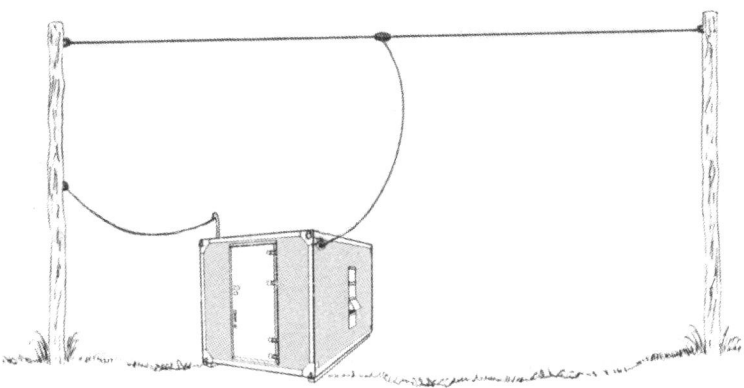

This diagram shows the older style beacon that uses a flat-top "T" antenna.

Transmitters such as the SR-505 manufactured by Scientific Radio Systems are commonly used.

Listening to Longwave　　　　　　　　　　　　　　　　　　　　Page 11

Because of the secure nature of most beacon sites, the only things most onlookers see are the shelter and the antenna system. Inside the shelter is the beacon transmitter and an antenna tuning unit (ATU). Since there is on one on premises most of the time, beacon transmitters are very robustly designed, and sometimes include a reserve unit that comes on-line in the event the primary transmitter fails. Most modern beacon transmitters also include alarm-reporting circuitry to alert maintenance personnel to problems.

The ATU's job is to provide a suitable electrical, or impedance, match between the transmitter and the station antenna. Although beacon antennas may appear to be rather large at first glance, they are (with few exceptions) actually far too short to work efficiently on the low frequencies. A tuning network compensates for the inadequacies of a mismatched, electrically short antenna. Some ATUs automatically re-tune as necessary to maintain a proper match when weather conditions change.

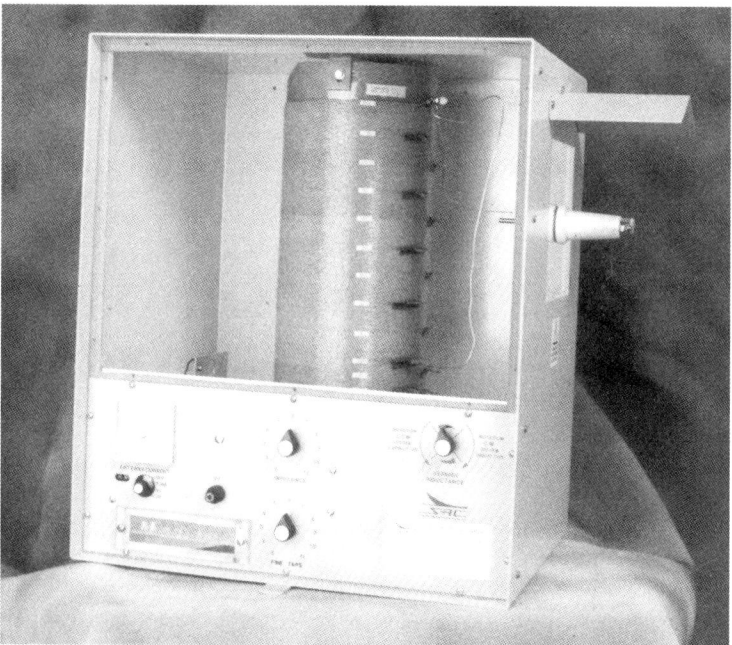

Antenna tuning units such as this one by Southern Avionics Co. provide an impedance match between a beacon transmitter and its antenna system. (Photo by Southern Avionics Co.)

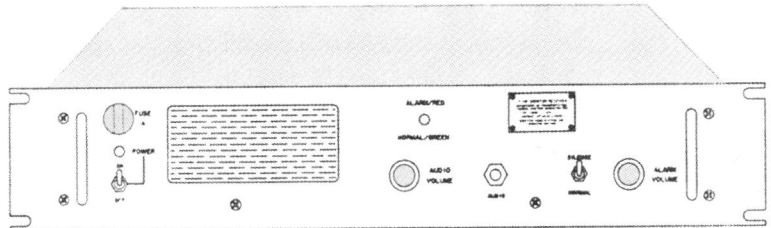

LF/MF Alarm Receivers such as this one are used at airports to monitor beacon performance. They provide a visual and aural alarm for loss of carrier or modulation, or if the tone is no longer interrupted.

Beacons, also known as NDBs (Non-Directional Beacons), are popular game for longwave listeners. Many beacon-hunters have logged hundreds. There are thousands in use worldwide, and as noted on the chart of the longwave spectrum, most will be found between 190 and 530 kHz.

Generally speaking, NDBs run fairly low power. Those in the United States average 25 watts or so. A handful of beacons used by the Federal Aviation Administration to transmit voice (AM) weather forecasts along with their Morse identifier are exceptions. They often operate at 500 to 1,000 watts.

Canadian beacons are another story, averaging several hundred watts. Many run 1,000 to 2,000 watts. As you'd expect, they can be heard at greater distances than their U.S. counterparts.

Other differences between U.S. and Canadian beacons can help you recognize the type to which you are listening. Most U.S. NDBs continuously repeat their identifiers. In Canada, the identifier is followed by a long dash. Many Canadian stations begin their call with a "Y." And quite a few include numbers; most U.S. beacons use only letters.

Some of the most widely-heard NDBs are listed on the next page. This list, and other lists of station frequencies used in this book are purposely limited in length. This is because the status of the radio spectrum is very fluid – new stations appear; old ones are retired; others change frequency. Only the most-established and strongest stations are shown.

Selected Non-Directional Beacons

198	DIW	Dixon, North Carolina
203	DMZ	Dickson, Tennessee
206	VNC	Venice, Florida
206	GLS	Galveston, Texas
212	ESN	Easton, Maryland
215	UIZ	Utica, Michigan
216	CLB	Wilmington, North Carolina
217	HZD	Huntingdon, Tennessee
230	PD	Pendleton, Oregon
230	NRN	Norton, Kansas
236	FOR	Forsyth, Montana
239	HKF	Middletown, Ohio
241	PVG	Portsmouth, Virginia
245	NKT	Cherry Point, North Carolina
245	YZE	Gore Bay, Ontario
246	DFI	Defiance, Ohio
247	ILT	Isleta, New Mexico
248	FRT	Spartansburg, South Carolina
248	KZ	Toronto, Ontario
253	UR	Burbank, California
257	SQT	Melbourne, Florida
258	ORJ	Corry, Pennsylvania
260	JH	Jackson, Mississippi
260	YSQ	Atlin, British Columbia
263	GR	Grand Rapids, Michigan
263	YGK	Kingston, Ontario
269	SWT	Seward, Nebraska
272	TYC	Campbellsville, Kentucky
272	YQA	Muskoka, Ontario
275	GEY	Greybull, Wyoming
276	TWT	Sturgis, Kentucky
276	YHR	Chevery, Quebec
277	ACE	Kachemak, Alaska
278	FD	Poplar Bluff, Missouri
278	CRZ	Corning, Iowa

281	EWK	Newton, Kansas
281	TOT	Denver, Colorado
284	DPG	Dugway Proving Grounds, Utah
296	ARF	Albertville, Alabama
296	LQR	Larned, Kansas
299	TR	Bristol, Tennessee
299	HW	Wilmington, Ohio
305	RO	Roswell, New Mexico
320	OM	Omaha, Nebraska
329	CH	Charleston, South Carolina
329	PJ	Whitehorse, Yukon Territory
332	HK	Chicago, Illinois
337	NA	Santa Ana, California
338	POB	Fayetteville, North Carolina
338	CMQ	Anchorage, Alaska
341	EGV	Eagle River, Wisconsin
344	AVN	Avon, New York
344	CL	Cleveland, Ohio
344	FCH	Fresno, California
346	LW	Lewisburg, West Virginia
347	AJR	Cornelia, Georgia
350	CWH	Huntsville, Alabama
350	RG	Oklahoma City, Oklahoma
350	ME	Chicago, Illinois
351	SI	Covington, Kentucky
352	QG	Windsor, Ontario
353	LI	Little Rock, Arkansas
355	CGE	Cambridge, Maryland
356	FOX	Fairbanks, Alaska
359	BO	Boise, Idaho
362	LYL	Lima, Ohio
366	YMW	Maniwaki, Qubec
368	IMR	Marshfield, Massachusetts
371	FND	Baltimore, Maryland
379	BRA	Asheville, North Carolina

Listening to Longwave

382	CR	Corpus Christi, Texas
388	CDX	Somerset, Kentucky
392	BAJ	Sterling, Colorado
394	ENZ	Nogales, Arizona
394	EZZ	Cameron, Missouri
400	FGX	Flemingberg, Kentucky
400	RO	Rochester, New York
404	LVV	Lake Lawn, Wisconsin
404	XCR	Little Falls, Minnesota
411	HDL	Holdenville, Oklahoma
413	CBC	Anahuac, Texas
413	YHD	Dryden, Ontario
413	OEG	Yuma, Arizona
417	HHG	Huntington, Indiana
432	MHP	Metter, Georgia

Besides keeping a reception log, "beacon hunters" often collect QSL (confirmation) cards from the operators of the beacons they hear. Since beacons fall into the "utility" category of stations (meaning they are not conventional broadcast stations), the operators of these stations rarely have QSL cards available. Listeners ordinarily create their own cards and mail them to the station engineer for a verification signature. When the QSL card is accompanied by a polite letter and a self-addressed, stamped envelope, most hobbyists report an excellent success rate. QSL requests can frequently be sent to the air facility nearest the beacon site. Check the reference section of this book for a list of beacon directories containing airport information.

Collecting QSL cards is a favorite activity of beacon chasers. These homemade cards brought results when mailed to the Engineer-in-Charge (along with a polite note and self-addressed stamped envelope).

DGPS BEACONS

The satellite-based Global Positioning System (GPS) has become extremely popular since its introduction to the public in the early 1990s. It is used for precise military and civilian navigation, electronic map displays in cars, automated survey work and instrument landing systems in some aircraft. A small handheld GPS receiver can now be purchased for under $200.

Starting in 1993, some Coast Guard beacons were retrofitted for a new service called *Differential* GPS (DGPS), to improve the accuracy of GPS via correction signals sent by longwave beacons.

Before DGPS, civilian GPS users could expect an accuracy of 100 meters (about 330 feet), which is not bad. But DGPS is 10 times better with an accuracy of less than 10 meters (30 feet). That is important to shipping interests operating in harbors or other congested areas.

How does DGPS accomplish this amazing precision? The exact latitude and longitude coordinates of radio beacons are known, so the error in the received GPS signal can be analyzed at the beacon site and the appropriate correction signals generated. These corrections are then transmitted to other users within the vicinity of the beacon. Longwave's ground-hugging characteristics make it ideal for this task.

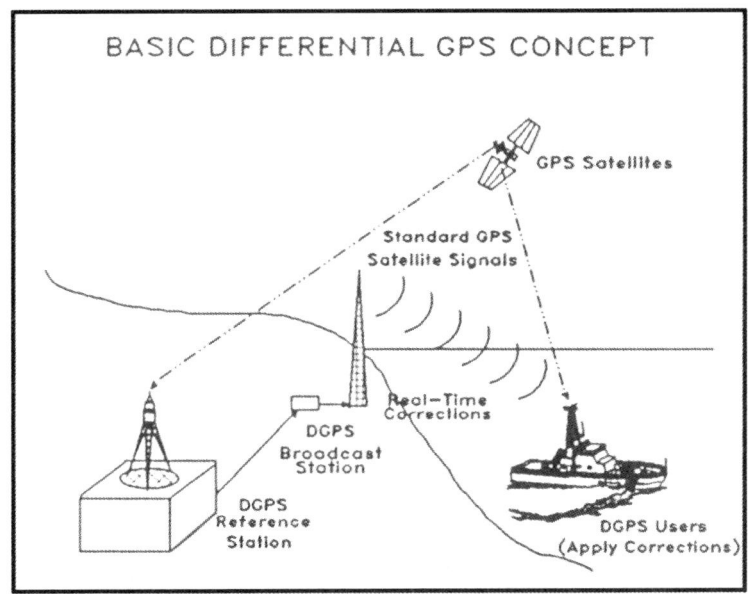

DGPS beacons greatly improve the accuracy of GPS within a local area.

You will know a DGPS beacon when you hear it. These stations send an Audio Frequency Shift Keying (AFSK) data stream, which produces a warbling note similar to narrow-shift radioteletype (RTTY). The mode is officially known as Minimum Shift Keying (MSK).

Originally most DGPS signals were sent "piggyback" on top of the regular Morse ID of the host beacon, the only effect being a warbled keyed tone. Today however, the Coast Guard operates these beacons with continuous DGPS data. Coast Guard surveys showed that few people used the beacons for direction finding anyway, so the keyed station identifier was no longer necessary.

Around the same time, the Coast Guard decommissioned the majority of its coastal radio beacons, including the sequenced networks around the Great Lakes. Those retained for DGPS service were higher powered or in strategic locations offering maximum coverage. The DGPS retrofit also included more efficient antennas at some sites, so most listeners should be able to hear at least a few of them.

A recent development in the DGPS story involves former Ground Wave Emergency Network (GWEN) stations. These operated at 150-175 kHz and sent heavily encrypted data between various Air Force sites around the U.S. GWEN stations were intended to provide "survivable" communications in the event of a nuclear war.

With the end of the Cold War, many such stations were taken off the air and mothballed. The Federal government then selected several GWEN stations to be retrofitted for DGPS in the 285-325 kHz band. So listeners may hear several new, much louder signals popping up in the DGPS band.

The new stations are part of the U.S. Department of Transportation's Nationwide Differential GPS (NDGPS) program. These sites use 300 foot "hot" towers to provide signal ranges of approximately 250 miles. Additional sites are expected to come on the air as funding, property, and environmental concerns are met.

Selected DGPS Stations in the U. S. (frequencies in kHz)

East Coast
Brunswick, ME (316)
Cape Canaveral, FL (289)
Cape Henlopen, DE (298)
Cape Henry, VA (289)
Charleston, SC (298)
Chatham, MA (325)
Fort Macon, NC (294)
Isabela, Puerto Rico (295)
Key West, FL (286)
Miami, FL (322)
Montauk Point, NY (293)
Portsmouth Hbr, NH (288)
Sandy Hook, NJ (286)

Gulf Coast
Aransas Pass, TX (304)
Egmont Key, FL (312)
English Turn, LA (293)
Galveston, TX (296)
Millers Ferry, AL (320)
Mobile Point, AL (300)

West Rivers
Memphis, TN (310)
Kansas City, MO (305)
Rock Island, IL (311)
Sallisaw, OK (299)
St. Louis, MO (322)
Vicksburg, MS (313)

Great Lakes
Cheboygan, MI (292)
Detroit, MI (319)
Milwaukee, WI (297)
Neebish Island, MI (309)
Saginaw Bay, MI (301)
St. Paul, MN (317)
Sturgeon Bay, WI (322)
Upper Keweenaw, MI (298)
Whitefish Point, MI (318)
Wisconsin Point, WI (296)
Youngstown, NY (322)

West Coast
Annette Island, AK (323)
Appleton, WA (300)
Cape Hinchinbrook, AK (292)
Cape Mendocino, CA (292)
Cold Bay, AK (289)
Fort Stevens, OR (287)
Gustavus, AK (288)
Kenai, AK (310)
Kodiak, AK (313)
Kokole Point, HI (300)
Pigeon Point, CA (287)
Point Arguello, CA (321)
Point Blunt, CA (310)
Point Loma, CA (302)
Potato Point, AK (298)
Robinson Point, WA (323)
Whidbey Island, WA (302)
Upolo Point, HI (286)

Listening to Longwave

MARITIME CW

The upper end of the longwave spectrum was once used for ship-to-shore communications in Morse code. Some traffic is still here, but with the advent of satellite systems such as INMARSAT and GMDSS, two-way maritime traffic has become rare. At this writing, CW weather broadcasts from shore stations are still be heard with some regularity. Of course, the closer to shore points one lives, the more activity can be heard. Nighttime reception is much better than during the day.

One place to listen for maritime traffic is 500 kHz, formerly the international distress and calling frequency. Usually this point on the dial is used to establish initial contact between stations, after which the parties involved move to another frequency. Most nations have discontinued full-time monitoring of 500 kHz. The frequency's special status may eventually be abolished internationally.

Despite the trend to abandon longwave maritime Morse Code communications, the Federal Communications Commission (F.C.C.) in 2005 granted a license for a new common-carrier, class 1A CW coast station to the Maritime Radio Historical Society (MRHS), the group that brought ex-RCA coast station KPH back to life. The MRHS applied for the license to assure that U.S. commercial Morse operations will continue into the future. The callsign of the station is KSM, located in San Francisco, California. It operates on longwave frequencies of 426 and 500 kHz (and shortwave frequencies of 6474 and 12993 kHz) with 5 kW power on all frequencies.

A still even more interesting idea has been proposed, namely the creation of a 600 meter amateur band. The recommended frequency allocation would be 495 to 510 kHz. This amateur band would offer unique propagational characteristics. Extremely reliable regional communications, based on ground-wave communications could be established. This could potentially provide uninterruptable emergency and security communications.

NAVTEX STATIONS

A prominent exception to the drop in marine traffic can be found at 518 kHz. This is the home of NAVTEX, an internationally standardized method of sending bulletins to boats equipped with low-cost digital receiving equipment. NAVTEX capability is now *required* for large vessels as part of the Safety of Life at Sea (SOLAS) convention, as amended in 1988.

The bulletins are primarily intended for waters up to 200 miles from shore and contain information about radio navigation status, search-and-rescue operations, weather forecasts, minesweeping drills, and other pertinent data.

NAVTEX bulletins can be easily read by hobby listeners using simple equipment. The first consideration for reading NAVTEX is the receiver itself. It's best if it has an RTTY mode to optimize the bandwidth for NAVTEX tones. However, any stable receiver with an SSB/CW setting or a BFO (Beat Frequency Oscillator) should provide satisfactory results.

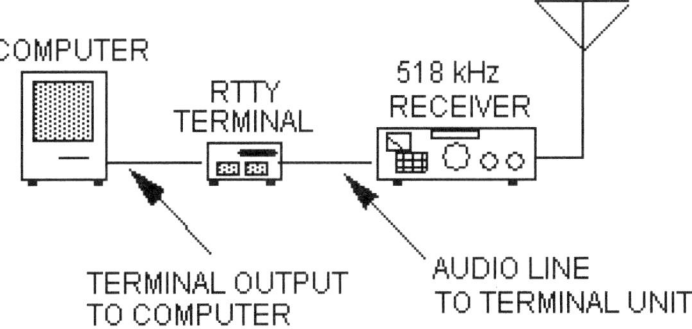

Typical setup for NAVTEX reception (hardware-based).

The additional equipment required for NAVTEX is a personal computer and a terminal unit (TU), running the appropriate software. Self-contained "readers" that include a built-in display screen are available. Readers connect directly to the receiver's audio output and often include a printer port to provide hard copy of intercepts.

Specialized software is another increasingly popular way to decode NAVTEX and other digital modes. It is no longer necessary to use a hardware-based terminal unit, provided you have a PC with a sound card and appropriate software. The PC processes the audio signals heard, and displays the results on screen. PC-based decoding is hard to beat for simplicity and ease of installation. All that is required is an audio cable between your PC and your radio receiver. Software controls are used to select the desired mode, audio levels, bandwidths, and other parameters.

NAVTEX is transmitted in SITOR Mode B (FEC Mode). This is similar to the AMTOR protocol used by amateur radio operators, In fact, most ham-grade RTTY terminal units can receive NAVTEX by simply selecting "AMTOR Mode B." But it is meant for one-way broadcast as opposed to the "chirp-chirp" two-way exchanges commonly heard on the ham bands.

```
19-Feb-92  00:57:55  ZCZC QA79
CCGD11 BNM 0202-92
1. CA-SEACOAST-GULF OF SANTA CATALINA
A POSSIBLE HOUSE TRAILER, CYLINDRICAL, APPROXIMATLEY
15 FEET LONG AND 8 FEET IN DIAMETER, GRAY IN COLOR WITH
A SPEAKER IN THE FRONT, MATTRESSES, PONTOONS AND THE
WORDS WAY TOO HIP PAINTED ON THE SIDE WAS SIGHTED IN
APPOXIMATE POSITION 33-12.6N 118-04.2W. MARINERS ARE
REQUESTED TO USE CAUTION WHEN TRANSITING THE AREA.
NNNN
```

NAVTEX messages such as this one can be received on 518 kHz with simple monitoring equipment.

Selected U.S. NAVTEX Stations (518 kHz)

Location	Transmission Times (UTC)[1]
Boston, Massachusetts	0445, 1045, 1645, 2245
Portsmouth, Virginia	0130, 0730, 1330, 1930
Miami, Florida	0000, 0600, 1200, 1800
San Juan, Puerto Rico	0415, 1015, 1615, 2215
New Orleans, Louisiana	0300, 0900, 1500, 2200
Long Beach, California	0445, 1045, 1645, 2245
San Francisco, California	0400, 1000, 1600, 2200
Astoria, Oregon	0130, 0730, 1330, 1930
Kodiak, Alaska	0300, 0900, 1500, 2115
Adak, Alaska	0000, 0500, 1200, 1745
Honolulu, Hawaii	0040, 0640, 1240, 1840
Guam	0100, 0700, 1300, 1900

As previously mentioned, software-based decoding has become increasingly popular for digital modes. The following page illustrates some hardware-based units that are still be used by some listeners. They often can be found secondhand at hamfests and swapmeets.

---

[1] The times shown in this chart are in UTC or coordinated universal time. This was formerly known as Greenwich mean time (GMT). If you own a shortwave radio you can hear the latest time in UTC, by tuning WWV, Fort Collins, Colorado or WWVH, Kauai, Hawaii on 2500, 5000, 10000 and 15000 kHz. You can also calculate the time yourself. From Eastern Standard time add 5 hours (4 PM = 2100 UTC). From Eastern Daylight time add 4 hours. From Central Standard add 6 hours, from Central Daylight add 5 hours, etc.

Listening to Longwave

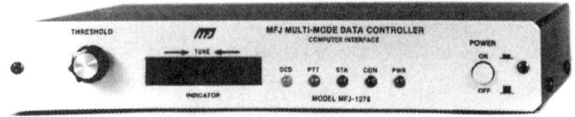

The MFJ-1278B is a full-featured demodulator that handles NAVTEX and many other digital modes such as CW, FAX, Packet, and SSTV.

The self-contained Universal M-450 Reader provides reception of NAVTEX and many other digital modes. Although a computer is not required, this unit contains an RS-232 port for PC interface.

The AOR ARD2 is a self-contained reader for ACARS and NAVTEX. It provides portable operation from four internal AA cells or external 12 VDC. A computer is not required. It supports NAVTEX-E for English and NAVTEX-J for the Japanese character set, used in coastal waters around Japan. Incoming text is displayed on a 16 character, 2 line LCD screen. A built-in speaker with volume control allows audio monitoring.

STANDARD FREQUENCY AND TIME STATIONS

Standard frequency and time stations transmit a continuous signal on a given precise frequency, with pulses at exact time intervals. The purpose of these stations is to provide extremely accurate time and frequency information for scientific purposes and for the general public.

Who utilizes such information? Scientists, astronomers, utility companies (to maintain the 60 Hz AC line frequency), the aerospace industry, aircraft and ship navigation systems, spacecraft guidance systems, and anyone interested in the correct time — the military, radio and TV stations, transportation companies and more.

Additional services also provided by some standard frequency and time stations include:
- precise audio frequency tones
- radio propagation forecasts
- geophysical alerts
- information on solar activity and marine storm warnings.

For North American listeners, one of the strongest signals that can be heard below 100 kHz is WWVB (60 kHz Fort Collins, Colorado). It can be identified by its pulsating carrier that loosely resembles Morse Code. WWVB is a sister station to well-known time stations WWV and WWVH that can be heard on 2.5, 5, 10, 15 and 20 MHz. It is run by the National Institute of Standards and Technology (formerly the National Bureau of Standards) in Fort Collins.

Like its HF counterparts, WWVB transmits highly accurate time signals, but omits voice announcements. Rather, WWVB transmits encoded data that can be interpreted by specialized receivers, test equipment, and some consumer-grade clocks.

WWVB's signal is also used as a frequency standard by many laboratories and power utilities to calibrate their equipment to exacting standards. Although some of these functions could be carried out on shortwave frequencies, variations in HF propagation can introduce delays or distortions that would be unacceptable to these users. Low-frequency signals travel primarily by ground wave and tend to be very stable.

WWVB maintains the accuracy of its frequency to better than 1 part in 100 billion and deviations from day to day are normally less than 1 part in 1,000 billion! How is this done? By the use of cesium-beam atomic clocks. The operation of these devices is based on the "atomic second", defined as 9,192,631,770 oscillations of an atom of cesium-133. Needless to say, matters could become very involved if this subject were carried further, but there is no need for that. The end result is a time standard so accurate that clocks of this nature gain or lose no more than several seconds in 100,000 years! Hardly a worry for the average individual who simply wishes to know the correct time.

Starting in 1997, an upgrade at WWVB raised the output power, lengthened the antenna system and added a new backup generator and power supply system. Three refurbished U.S. Navy transmitters were placed in service at WWVB to provide the main signals, as well as a backup capability.

WWVB's power output went from 13 kW to nearly 50 kW. Combined with the antenna changes and other improvements, this provides excellent signal strength throughout most of the U.S. and Alaska. This clears the way for expanded use of WWVB in many consumer timekeeping applications such as VCRs, wall clocks and even wrist watches.

WWVB QSL card from NIST.

Those interested in additional information on the NIST and the services it provides can write to the National Institute of Science & Technology, 2000 East County Road 58, Fort Collins, CO 80524.

Inexpensive consumer clocks such as these units by LaCrosse Technology and MFJ use time signals from WWVB, 60 kHz, to maintain exceptional accuracy!

FOREIGN BROADCASTS

Between 148.5 and 283.5 kHz, AM broadcast stations from Europe, Asia and Africa can be heard. Many are quite powerful, running from half a million to several million watts. Residents of the East and West coasts are most likely to hear these transmissions, while catches for those in central states are more difficult. The key to successful reception is to have a path of darkness between you and the transmitting station. Winter nights are best for listening because of reduced levels of natural static (QRN). If your receiver allows for programming preset "channels," entering the frequencies of several broadcasters into memory may be helpful. In this way, a quick check can be made for propagation quality. Note that VLF broadcasters do not use call signs, so the list provided here shows locations only.

The *World Radio TV Handbook* is *the* reference source for this type of broadcaster. It lists frequencies, locations and power outputs of all international longwave stations.

Selected Longwave Broadcast Stations

Frequency kHz	Location	Power kW
153	Mainflingen, Germany	250
153	Béchar, Algeria	1000
162	Allouis, France	2000
171	Tbilisskaya, Russia	1200
177	Berlin, Germany	500
180	Polati, Turkey	1200
183	Saarlouis, Germany	2000
189	Reykjavik, Iceland	300
198	England (various locations)	up to 500
207	Aholming, Germany	250
216	Roumoules, France	1400
225	Racz'yn, Poland	1000
234	Beidweiler, Luxemburg	2000
243	Kalundborg, Denmark	300
252	Clarkstown, Ireland	500
261	Taldom, Russia	2500
270	Topoina, Czech Republic	650

Listening to Longwave

Chapter 5
Special Users of Longwave

As one goes down in frequency, unusual things begin to happen. The surface of the Earth and the lowest layer of the ionosphere work as a kind of pipe or guide, that carries radio waves with little signal loss. And materials that stop higher-frequency radio waves cold can be penetrated by waves near the bottom of the spectrum. Earth and water are no barrier. Suddenly communications underground, or through water, and reliability under adverse conditions where other frequencies fail, becomes possible. It's a whole new ball game – and an exciting one!

But nothing comes free. Let's pick a frequency at random – say 5 kHz. A half-wave dipole antenna for this frequency would be 18 miles long! Data transfer is another problem. One could hardly use this frequency for listening to a favorite symphony. The specs on high fidelity speakers might show a frequency range of 50 to 20,000 Hz, or nearly 20 kHz. That's wider than the entire band in which the transmission is taking place! Like trying to transmit a 6 MHz wide TV signal in the 200 kHz bandwidth allowed between FM radio stations. It can't be done.

Despite such drawbacks, areas near the bottom of the longwave spectrum are utilized, but it requires huge antennas, enormous power levels and limited types of emission. The story of users way down under makes an interesting tale. Following are accounts of some of those who utilize this portion of the radio spectrum.

LORAN

LORAN is a navigation system used principally by ships and planes to accurately determine position. Special receivers on board pick up signals from powerful transmitters around the world and analyze minute differences in the time transmissions arrival from multiple stations. The receivers then determine hyperbolic lines of position, which locate the craft in question to within a quarter of a mile. In the illustration below, a craft in the Atlantic is determining a position via LORAN signals coming from Florida and North Carolina.

Once position is determined, LORAN can return a ship or plane to the same spot with a repeatability accuracy of 50 meters (164 feet). Such precision enables the system to be used as a reference marker for pinpointing key targets of interest such as fishing spots and capped oil wells.

LORAN signals are transmitted between 90 kHz and 110 kHz with a center frequency of 100 kHz. The sound might best be described as "pulsed, wideband" static or "clickety-clack" static. Once you hear it, you will know you have found it.

Statistics on LORAN transmitting sites are impressive. Power levels range from 300 kilowatts to 2 megawatts. This translates into expensive operating costs. For example, a former station on Nantucket Island averaged electricity fees of $8,000 per month in 1984.

A typical antenna consists of a 625-foot vertical with a capacitive top hat. Some antennas rise to heights of 1,400 feet. Such installations, in addition to making effective radiators, also act as giant lightning rods. As a result, stringent precautions must be taken to protect lives and property. One common method of lightning protection makes use of a Z-shaped lead-in from the antenna to the building housing the transmitting gear. The lower part of the "Z" is brought near ground potential so a lightning strike will arc across to ground rather than follow a path which would lead to certain destruction of equipment inside.

Plans to terminate LORAN operations on December 31, 2000 were announced in the 1996 Federal Radio Navigation Plan. However, a congressionally mandated report on the future use and funding of LORAN, prepared for Department of Transportation (DOT) in 1998, recommended LORAN operations continue beyond year 2000. As of Fall 2006, the decision to terminate operations was currently under review by the DOT.

Whatever the date, LORAN's days are probably numbered. There have been some attempts to enhance and revitalize LORAN, but GPS offers better accuracy, simplified operation and compact receiving equipment. LORAN has provided faithful service for over fifty years, but it appears the time has finally come to yield to progress.

ALPHA

The Russian-based ALPHA system is a radio navigation network that works by measuring small but significant phase differences from three transmitter sites across Russia. Operating at 11.905, 12.649, and 14.881 kHz, ALPHA signals can be easily identified as they consist of short (0.4 second) tones of different pitch. The transmission cycle of the ALPHA system is repeated every 3.6 seconds.

Depending on your location, you may or may not be able to hear all three ALPHA stations. In North America, those living near the west coast will probably hear the station near Khabarovsk, (far eastern Russia) the loudest. For those on the east coast of North America, the station at Krasnodar (Black Sea) will likely be the strongest. The remaining ALPHA station is at Novosibirsk (central Russia). It is usually weak or unreadable by North American listeners.

Until 1997, a similar navigation system near 12 kHz was operated by the U.S in cooperation with several other nations. The system was called OMEGA and could be identified by its melody of somewhat longer tones than those used by ALPHA. The following provides some historical information on OMEGA.

OMEGA

OMEGA was a radio navigation system that measured the phase differences between signals from various transmitters. Worldwide coverage was provided by only eight transmitting sites located in Argentina, Australia, France, Japan, Hawaii, Liberia, Norway and North Dakota. A positioning accuracy of 2 to 4 nautical miles was possible with this system, which utilized frequencies between 10 and 14 kHz. OMEGA's signals were easily spotted as strange-sounding tones of different pitch.

As might be expected from descriptions of other installations, "big" best described equipment associated with this system. For example, the antenna at Tsu-Shima, Japan was 10 feet wide and 1,500 feet high – the tallest structure in the Orient. It was anchored by guy wires to surrounding islands. In Norway, the antenna array spanned a fiord. And in Hawaii, a volcano crater was crisscrossed with the massive antenna array there.

An interesting aspect of OMEGA was that its useful range was limited not by signal strength, but by interference from its own signal circling the globe to arrive back at its origin.

GWEN

GWEN stands for Ground Wave Emergency Network. GWEN was a system of VLF transmitting sites scattered throughout the United States for the purpose of providing communications in the event of a nuclear attack. It was developed by the U.S. Air Force in cooperation with the Federal Emergency Management Administration (FEMA). This project was terminated in 1998.

NAA

NAA is a radio station located in Cutler, Maine which is operated by the U.S. Navy as a reliable communications link between high level military command on shore and ships, planes and submarines operating in the northern latitudes. It has a sister station, NLK, located in Jim Creek, Washington which provides the same services for other parts of the globe. NAA operates on 24.0 kHz and NLK on 24.8 kHz.

Facts and figures concerning NAA are nothing short of astounding. To appreciate the enormity of this operation, consider that:

— NAA is situated on 3,000 acres surrounded on three sides by water.

— The compound has 12 miles of roads.

— Construction required 90,000 cubic yards of concrete and 15,000 tons of structural steel.

— It has its own power plant, utilizing four Cooper-Bessemer and other generators with combined capacity of 15 megawatts (15,000,000 watts).

— Its reserve fuel capacity is 20,000 barrels of oil. A one-penny-per-gallon price change adds or subtracts $5,000 or more from the cost of filling its tanks.

— Its a power output is 2 mega watts.

— After leaving the transmitter, the radio signal is split and sent through two half-mile-long tunnels to helix houses and loaded into two antenna arrays, each a base-fed monopole with a top hat for capacitance tuning.

— Each array utilizes 13 towers, a center tower of 980 feet ringed by six of 875 feet and six of 800 feet. The tallest of these is taller than the Washington Monument and Bunker Hill combined.

— The arrays use 396,000 feet of one inch phosphor bronze wire above ground plus another 13 million feet for grounding. The ground wires are terminated in the surrounding waters and secured by 250 sea anchors.

— The antenna system occupies 715 acres, large enough to house 22 Pentagon buildings.

— With 75 miles of wire in the air, freezing rain builds up of tons of ice. To counteract this problem, the radio signal is temporarily fed to one antenna only and up to 9,000,000 watts of current is applied to the other antenna to heat its wires and melt the ice. When it is free of ice, the signal is switched to that array and the other antenna is cleared.

— To keep ice build-up from damaging the arrays, a gravity counterweight system automatically allows the antennas to lower or rise. The counterbalance system consists of 36 200 foot towers, which serve as giant tracks to guide counterweight drums filled with dense concrete, each weighing 220 tons.

PROJECT ELF

One of the grandest schemes for VLF communications ever devised, and one that literally and figuratively scraped the bottom, was Project ELF, a $240 million system for communicating with submarines.

Despite numerous legal battles by environmentalists and vandalism from opponents to the system, Project ELF was operational from two sites in the north-central United States. A 56-mile long antenna was installed in the Escanaba River State Forest in Michigan's Upper Peninsula, atop 55-foot poles along a swatch 50 to 100 feet wide cut through the woods. In Clam Lake, Wisconsin, a similar 28-mile long antenna was erected.

General Telephone and Electronics, the major contractor, constructed a system transmitting on 76 Hertz (yes, 76 cycles per second) to send messages to submarines at depths of 300 to 400 feet. Because of the extremely low frequency, data had to be transmitted at very low rates. Sending a single three-symbol message took 15 minutes!

Because such a small amount of data could be transmitted and there were no receiving facilities at the Michigan or Wisconsin sites, the main purpose of this project may well have been to send an initial-strike launch signal to our nuclear equipped submarines. Another possibility is that the system served as a "bell ringer" to alert sub crews to surface communication on shortwave (HF) or satellite frequencies. Project ELF terminated operations on September 30, 2004.

OTHER STATIONS BELOW 150 kHz

Listening below 150 kHz is different from monitoring the rest of the longwave band. Signals tend to follow ground wave paths – they do not rely heavily on skywave propagation to cover long distances. It is not uncommon to hear distant stations at nearly the same strength both day and night.

Stations below 150 kHz also rarely identify themselves in plain language. Most signals consist of usually encrypted data (FSK), or on rare occasions, CW (Morse Code). The majority of these stations are run by the military or government agencies, and are not required to identify at regular intervals.

The best way to identify these stations is by consulting a reliable frequency guide. There are very few shared frequency assignments in this range, so if you look up a signal on a given frequency, you can be quite certain of its identity. To get you started, the table on the next page lists some sub-150 kHz stations believed to be active as of this writing. Keep in mind that long periods may go by without any activity heard on a given frequency.

FREQ. (kHz.)	ID	CITY	REMARKS
14.88	—	VARIOUS LOC., RUSSIA	CW PULSES, ALPHA NAV.SYS.
14.5	HWU	LE BLANC, FRANCE	FSK, FRENCH NAVY
14.7	NPM	LUALUALEI, HI	FSK, US NAVY
15	UIK	VLADIVOSTOK, RUSSIA	CW
16.4	JXN	NOVIKEN, NORWAY	FSK, NORWEGIAN NAVY, 45 kW
16.8	FUB	PARIS, FRANCE	FSK, FRENCH NAVY, 23 kW
17.1	UMS	MOSCOW, RUSSIA	CW, FSK, RUSSIAN NAVY
17.2	SAQ	GRIMETON, SWEDEN	CW, LAST REMAINING ALEXANDERSON ALTERNATOR
17.4	NDT	VOSAMI, JAPAN	FSK, US NAVY
18.1	RDL	RUSSIA	FSK, CW, RUSSIAN NAVY
18.2	VTX3	VIJAYA NARAYANAM, INDIA	CW
18.5	DHO	W. RHAUDERFEHN, GERMANY	CW, GERMAN NAVY, 500 kW
19.6	GBZ	CRIGGION, WALES	FSK, ROYAL NAVY, 44 kW
20.27	ICV	TAVOLARA, ITALY	FSK, ITALIAN NAVY, 43 kW
21.37	GYA	LONDON, ENGLAND	FSK, ROYAL NAVY
21.4	NPM	LUALUALEI, HI	FSK, US NAVY, 630 kW
22.3	NWC	EXMOUTH, AUST	FSK, US NAVY, 1000 kW
23	UTR3	RUSSIA	TIME SIGNALS
23	UQC3	RUSSIA	TIME SIGNALS
21-23.6	—	TACAMO (AIRBORNE)	CW, FSK, USAF
23.4	NPM	LUALUALEI, HI	FSK, US NAVY, 630 kW
24.0	NAA	CUTLER, ME	FSK, US NAVY, 1000 kW
24.8	NLK	JIM CREEK, WA	FSK, US NAVY, 250 kW
25	UTR3	RUSSIA	TIME SIGNALS
25	UQC3	RUSSIA	TIME SIGNALS
25.1	UTR3	RUSSIA	TIME SIGNALS
25.5	UTR3	RUSSIA	TIME SIGNALS
25.5	UQC3	RUSSIA	TIME SIGNALS
26.1	NPG	SAN FRANSICO, CA	CW, US NAVY
26.1	—	TACAMO (AIRBORNE)	CW, FSK, USAF
27.1	—	TACAMO (AIRBORNE)	CW, FSK, USAF
28.5	NAU	AQUADA, PR	FSK, US NAVY, 100 kW
30.6	NPL	SAN DIEGO, CA	FSK, US NAVY
40	JG2AS	CHIBA, JAPAN	STANDARD TIME & FREQ., 10 kW
44	VHB	BELCONNEN, AUSTRALIA	FSK, AUSTRALIAN NAVY
46.25	DCF46	MAINFLINGIN, GERMANY	TIME SIGNALS
48.5	FXL	SILVER CREEK, NE	FSK, USAF
50	OMA	PRAGUE, CZECH REP.	STANDARD TIME & FREQ.
50	RTZ	IRKUTSK, RUSSIA	TIME SIGNALS
51.95	GYA	LONDON, ENGLAND	FSK, ROYAL NAVY
53.0	NPL	SAN DIEGO, CA	FSK, US NAVY
54	NAM	NORFOLK, VA	FSK, US NAVY
54.05	NBA	BALBOA, PANAMA	FSK, US NAVY
55.5	GXH	THURSO, SCOTLAND	FSK, ROYAL NAVY
56.5	NPG	DIXON, CA	FSK, US NAVY

FREQ. (kHz.)	ID	CITY	REMARKS
60	MSF	RUGBY, ENGLAND	STANDARD TIME & FREQUENCY
60	WWVB	FORT COLLINS, CO	N.I.S.T, TIME SIGNALS, 45 kW
61.55	—	NATO	CW, FSK
62.6	—	FRANCE	FSK, FRENCH NAVY
65.8	FUE	BREST, FRANCE	FSK, FRENCH NAVY
66.66	RBU	MOSCOW	TIME SIGNALS
68.0	GBY20	RUGBY, ENGLAND	FSK, ROYAL NAVY
68.9	XPH	THULE, GREENLAND	FSK, USAF
73.6	CFH	HALIFAX, NS	FSK, CANADIAN FORCES
75	HBG	PRANGINS, SWITZERLAND	STANDARD TIME & FREQ., 20 kW
76.2	CKN	VANCOUVER, BC	FSK, CANADIAN FORCES
77.2	NAM	DRIVER, VA	FSK, US NAVY
77.5	DCF77	MAINFLINGEN, GERMANY	STANDARD TIME & FREQUENCY
82.5	MKL	PETREAVIE, SCOTLAND	FSK, CW
83.8	FTA83	ST ANDRE DE COROY, FR	CW
91.15	FTA911	ST ANDRE DE COROY, FR	CW
93.9	FOU	TOULON, FRANCE	CW
100	—	VARIOUS LOC. WORLDWIDE	LORAN NAVIGATION SYS.
119.8	NPG	DIXON, CA	FSK, US NAVY
122.5	CFH	HALIFAX, NS	CANADIAN FORCES, FSK
124	CKN	VANCOUVER, BC	CW, CANADIAN NAVY
128.25	NPL	SAN DIEGO, CA	FSK, US NAVY
131.05	FUF	MARTINIQUE	CW, FRENCH NAVY
133.15	CFH	HALIFAX, NS	FSK, CANADIAN FORCES
135.95	NPG	DIXON, CA	FSK, US NAVY
136	—	VARIOUS LOC.	AMATEUR/EXPERIMENTER BAND
137	CFH	HALIFAX, NS	FSK, CANADIAN FORCES
146.1	NPM	LUALUALAI, HI	FSK, US NAVY
148.2	NPL	SAN DIEGO, CA	FSK, US NAVY

GBZ (19.6 kHz), Criggion, Wales, is an example of a high power VLF station.

Chapter 6
LOWFERS

Lowfers is shorthand for Low Frequency Experimental Radio Station. Lowfers operate between 160 and 190 kHz – called the 1750 meter band – without a license, a privilege granted by Part 15 of the FCC Rules and Regulations.

What do they transmit? When do they transmit? Who can hear them? What are their power limits and other restrictions? Low Frequency Experimental Radio Stations are operated as their name implies, by experimenters or hobbyists. They are limited by law to one watt of input power to the final amplifier and their antenna length, including transmission line, may not exceed 50 feet (15 meters). These are the primary restrictions; several others will be discussed shortly.

Most of these hardy souls operate their stations as beacons (using Morse code) – some 24 hours a day 365 days a year, others on weekends or at night, and others simply when the mood strikes them. Since no license is required, no callsign is assigned by the FCC. Therefore an identifier must be devised by the station owner. The most common IDs used are comprised of the owner's initials or the letters following the numeral in his or her amateur radio call (if the station owner is also a ham). Sometimes the address of the station is given after the callsign so listeners know where to send reception reports.

Although operated as beacons, Lowfers also make contacts among themselves. CW or data is the mode used most often, but SSB or AM are sometimes employed. Some Lowfers have experimented with Coherent CW (CCW). This a form of CW that uses a very accurate frequency source, such as WWVB, to synchronize the local oscillator frequencies at both the transmitting and receiving ends. The exact code speed of the transmitter is stipulated in advance. By knowing the code speed and having the oscillators at each end synchronized, the receiver can use a filter with very narrow bandwidth. This can improve signal strength by more than 20 dB.

Another promising mode for Lowfer work is Binary Phase Shift Keying (BPSK). BPSK is a form of data transmission that resembles the packet mode used by many radio amateurs. It allows transmission between personal computers even in the presence of weak signals or interference. A clear benefit of this mode is that a receiving operator can tune to the frequency of a known Lowfer station and let the system run overnight. The next day, the "copy" can be checked to see if there were any successful intercepts.

QRSS, or super-slow CW, has become very popular in recent years. By using very slow on/off keying, signals can be received with an extremely narrow bandwidth, thus boosting the signal-to-noise radio of intercepted transmissions. How slow is QRSS? One popular scheme called "QRSS30" takes 30 seconds to send one dot in Morse Code! Special software is required to decode these transmissions. At present, *Argo* and *Spectran* are the most common programs, and they may be freely download from the Internet. An online search for these terms will provide links to download information.

Anyone interested in this unique hobby *must* get to know the Longwave Club of America (LWCA) which was organized "to promote DXing and experimenting on frequencies below 550 kHz and activity on the 1750 meter band." It is a club of high and long standing. Many of its contributors provide excellent technical articles concerning Lowfer and other activity on the longwaves. See the Reference Section for the address of this club and other organizations, publications, and other sources of information mentioned in this book.

For many years, Ken Cornell's *Low and Medium Frequency Radio Scrapbook,* published in 1977, was a mainstay of Lowfer information — covering topics such as receivers, antennas, transmitters and accessories. Unfortunately, Cornell died in 1997, and even used copies of the book are hard to find. If you get your hand on one, it makes an excellent addition to any Lowfer bookshelf. A new publication, *The LF Experimenter's Source Book,* has become available in recent years through the efforts of the Radio Society of Great Britain (RSGB). The book is a compilation of LF articles that have appeared recently in radio magazines. While the focus is on European operations at 136 kHz, the topics presented can be easily applied to the 160-190 kHz band.

No one knows for sure how many Lowfers there are. They are most numerous in the Northeast and California, but stations are known to exist in Florida, Illinois, Oklahoma, Nevada, New Mexico and South Carolina. The numbers have increased appreciably in recent years and indications are they will continue to do so. Despite the low power limit, reception from 25 to 75 miles away is common, with ideal conditions allowing signals to be heard from several hundred miles away. On occasion, reports of signals of over 500 miles distance are recorded.

The list below shows some selected stations believed to be active at press time. A list of this sort can only be considered a snapshot in time. Stations come and go, and change frequency and IDs. The list should, however, provide a starting point for those wishing to hear a LOWFER station.

Selected Lowfer Stations (136 kHz and 160-190 kHz)

FREQ.	ID	LOCATION
137.422	WD2XES	Holden, Massachusetts
137.573	VY1JA	Whitehorse, Yukon Terr.
137.776	MP	London, Ontario
137.777	VO1NA	Torbay, New Foundland
137.777	VE7SL	Mayne Island, Brit. Col.
137.777	WD2XKO	Stanfield, North Carolina
137.778	TIL	Vancouver, Brit. Col.
137.779	WD2XGJ	Wayland, Massachusetts
137.781	WD2XNS	Burlington, Connecticut
161.200	YTN	Minneola, Florida
164.900	KLFB	Sunnyvale, California
166.500	XSR	Jefferson, Louisiana
169.900	R	Cotter, Arizona

FREQ.	ID	LOCATION
172.800	SWB	El Sobrante, California
175.000	D	Des Moines, Iowa
176.200	2J	East Windsor, New Jersey
181.167	IZJ	San Gabriel, California
182.200	BRO	Duluth, Minnesota
183.500	PLI	Burbank, California
183.544	MEL	San Jose, California
183.673	NW	Williams, Oregon
184.325	WMG	Pittsburg, Kansas
184.700	GNB	Hagerman, New Mexico
184.800	BK	Shell Lake, Wisconsin
184.900	7JE	Waddell, Arizona
185.185	FAW	Riverton, Utah
185.297	COV	S. Coffeyville, Oklahoma
185.297	TAG	Raymond, Maine
185.298	IP	Agricola, Mississippi
185.299	NC	Stanfield, North Carolina
185.301	WM	Andover, Massachusetts
185.301	TMO	Richmondville, New York
185.301	VD	Burlington, Connecticut
185.303	WD2XGI	St. Francis, Minnesota
186.410	HCN	Magdalena, New Mexico
186.450	JWS	Kernersville, North Carolina
186.700	LEK	Aitkin, Minnesota
186.940	BOB	Mahomet, Illinois
187.370	VO1NA	Torbay, New Foundland
187.752	NHVT	Charlestown, New Hampshire
188.100	YHO	Mason, Ohio
188.800	EAR	Saltford, Ontario
189.390	TH	Colts Neck, New Jersey
189.500	MO	Seneca, Missouri
189.659	NWNJ	Hainesville, New Jersey
189.800	RM	Duluth, Minnesota
189.950	WEB	Rosenberg, Texas

> **LOW FREQUENCY EXPERIMENTAL BEACON**
>
> **KRY**
>
> JOE SALOKA   CHARDON, OHIO 44024

> **ZWI** **178.6 kHz**
> **1750 METER BAND**
> Howard "Mort" Mortimer
> 7614 Homestead Drive, Baldwinsville, NY 13027-9408
>
> ONONDAGA COUNTY USA, Grid FN13
> 10 Miles NW of Syracuse, NY
> This Confirms Your Reception of Lowfer Beacon
> "ZWI" At 178.6 kHz on_____ At_____ UTC.
> XMTR- 1 Watt Palomar   Ant. 50 ft. Sloper
> Thank You For Your Report.
>
> WB2 ZWI
> Operator: Mort   Date: 3-30-96

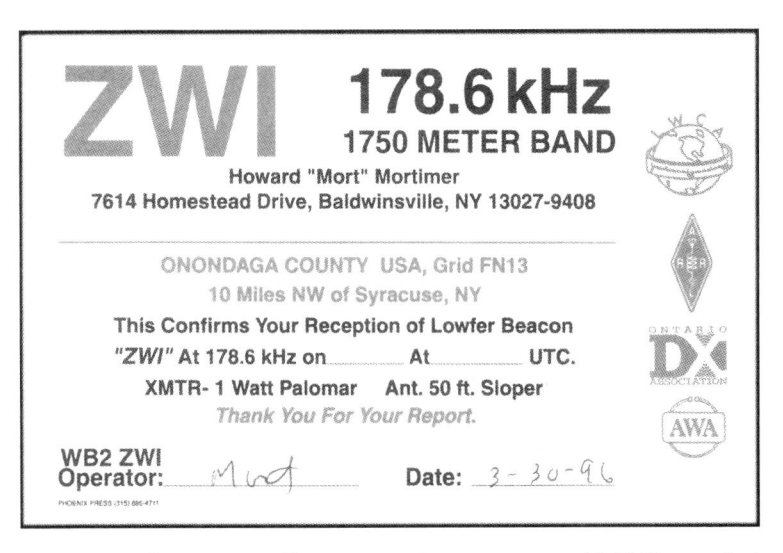

In many parts of Europe, radio amateurs have access to 136 kHz at relatively high power levels. (The actual frequency range is 135.7-137.8 kHz.) Although this operation is by licensed amateur stations, it deserves mention here because of its experimental nature. A similar allocation exists at 73 kHz for UK amateurs having a Notice of Variation from their government. The FCC is evaluating a proposal that would grant U.S. hams access to 136 kHz in addition to the 160-190 kHz band. Whatever LF allocation is given to hams, a debt of gratitude will be owed to the Lowfers who have carried out valuable experiments on these frequencies since the early 1970s using low power and very small antennas.

The equipment needed to transmit on longwave is relatively simple and inexpensive, and the rules are generally uncomplicated. Other than power level and antenna size limits, few other restrictions apply. Spurious emissions outside the 160 to 190 kHz segment must be suppressed by at least 20 dB and no interference may be caused to other services. The latter restriction is seldom a problem since the "other services" are likely to be CW or RTTY stations running much higher power!

To fully comply with FCC rules, one last step must be taken. The station owner "shall attach to each such device a signed and dated label that reads as follows":

**I have constructed this device for my own use.
I have tested it and certify that it complies with
the applicable regulations of FCC Rules, Part 15.
A copy of my measurements is in my possession
and is available for inspection.**

_____   _____
Signature   **Date**

The foregoing has been an introduction to the fascinating hobby that awaits anyone willing to give it a try. In the meantime, keep your ears peeled to the 30 kHz span between 160 and 190 kHz. Maybe you will find some Lowfers!

Chapter 7
Natural Longwave Signals

Strange things go on at the bottom of the radio spectrum. You may hear noises reminiscent of sound effects from old science fiction movies. At certain times, for example, unusual sounds can be heard between 3 and 15 kHz: clicks, pops, hisses, rushing and swishing sounds and whistles of descending pitch. Many of these noises are unexplained. The "whistlers" are believed to be caused by lighting.

Lightning has its greatest energy output at about 5 kHz, with descending intensity up and down the spectrum. Listening to lightning may seem somewhat ridiculous. After all, everyone has heard it in the form of static which carries through the longwaves, AM radio band, shortwaves, FM radio band and TV channels. At the very least it is a nuisance. If it strikes close to home it can be catastrophic. No one is suggesting you tune to 5 kHz when a thunderstorm is approaching!

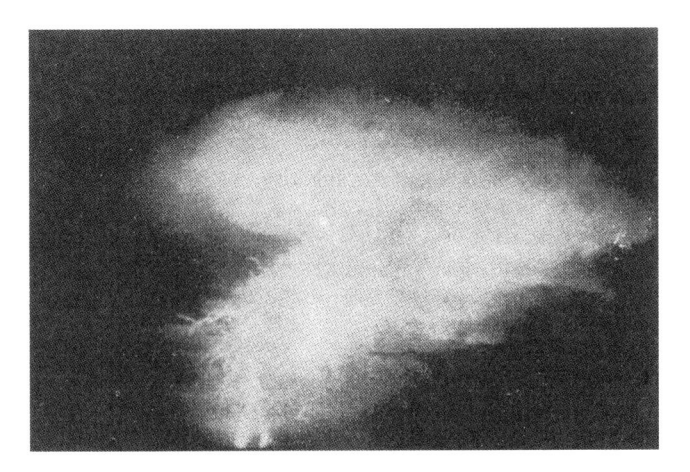

But it is distant lightning, according to the best current theories, that is responsible for the strange-sounding, descending whistles that can sometimes be heard during predawn hours. The other unusual sounds mentioned may be heard around the same time. Missile launches also generate radio wave emissions, but you'd have to be quite close to the launch site to monitor them.

In recent years, four of the most common natural radio sounds— *sferics, whistlers, tweeks* and *chorus*—have become better understood. The following provides a basic explanation and what you might expect to hear in the realm below 10 kHz.

Sferics (radio atmospherics)

These are the familiar "static crashes," clicks and pops that we all complain about when trying to snag a weak utility station. Although they are normally a nuisance, each sferic represents a lightning discharge, and their presence often indicates a greater probability of other natural radio sounds being heard, especially when they appear during a solar disturbance. Use common sense when listening to sferics. If they become *very* strong, it's best to shut down to avoid a lightning hazard!

Whistlers

Whistlers are the best known of all natural radio sounds. They produce a falling pitch that lasts from one to several seconds, depending on the distance the signal has traveled. The whistling note can range from a nearly pure tone to a coarse, "breathy" note.

It's well known that lightning produces whistlers, but it is the interaction between lightning, the Earth's surrounding magnetosphere, and the charged particles coming from the Sun (the solar wind) that give us the spectacular sound of a whistler.

Briefly, here's how happens: During a solar disturbance, the Earth's magnetosphere is bombarded by waves of charged particles. Besides setting off spectacular displays like the Aurora Borealis (Northern Lights) and Australis Borealis (Southern Lights), the charged particles can also create ionized trails or "ducts" along the magnetosphere's lines of force, allowing improved electrical conductivity.

The RF energy created can shoot through these ducts, traveling far into space, and ultimately returning to a conjugate point in the opposite hemisphere. Under the right conditions, a listener near this "landing zone" may hear the sound of a short, "one-hop" whistler.

A "two-hop" whistler occurs when the lightning's RF energy is reflected back into the ionized duct, returning to a spot near the originating stroke. This often results in a proportionately longer (but weaker) whistler. This flip-flopping may occur many times, producing progressively longer and weaker whistlers.

A mechanism called *dispersion* is responsible for a whistler's dropping note. Since higher frequencies travel slightly faster than lower ones, they reach the receiving station first, followed by progressively lower frequencies. The farther a whistler travels, the more pronounced the dispersion effect.

Tweeks

Tweeks are mostly a nighttime phenomenon occurring below 5 kHz. They occur when lightning's RF energy travels within a natural channel, or waveguide, formed between the Earth and the D and E layers of the ionosphere (approximately 40 to 70 miles above the Earth).

Tweeks produce a very short ping or chirp rarely longer than a fraction of a second. The cutoff point of these rapidly descending notes (usually around 1.5 kHz) represents the lowest frequency at which the dimensions of the waveguide can support the RF energy. (The dimensions of a waveguide must be more than a half wavelength of the RF energy to be carried.)

The dropping pitch of tweeks is caused by the same dispersion effect described for whistlers, except the whole process occurs *within* the ionosphere.

Chorus

The chorus phenomenon is named for its cacophony of overlapping squawks, whoops and chirps that *rise* in frequency. It sounds very similar to flocks of birds singing at sunrise—so much so that episodes are often called "dawn chorus." Chorus signals are believed to be caused by pulsations in the magnetosphere, that surrounds the Earth, during very active solar storms. Often chorus signals will come in distinct waves, rising and falling in intensity over the period of just a few seconds. These are known as "chorus trains."

Chorus events are somewhat rare. The best time to listen for them is generally during a solar storm in the early morning hours. It can also occur at night, especially when there is a visible aurora (Northern or Southern Lights) over the poles. As with most natural radio signals, the closer you are to the North or South Pole, the more frequent and intense the chorus activity will be.

Natural Radio Receiving Equipment

Very few communications receivers are designed to operate below 10 kHz. For this reason, those interested in natural radio most often homebrew their own receiving equipment, or purchase a commercial receiver that does operate at these low frequencies. Fortunately, the receiving circuits are uncomplicated, and the equipment is modestly priced as compared to the cost of a full-blown communications receiver.

In its simplest form, a natural radio receiver consists of a high-gain audio amplifier connected to an antenna. Earphones are connected to the amplifier output, and that's all there is to it. Interestingly, World War I soldiers may have been among the first to hear whistlers when they employed high-gain amplifiers to eavesdrop on field telephone communications. Upon hearing the falling pitches of whistlers, some operators believed they were hearing the grenades fly at the battlefront!

World War I soldiers were among the first to hear whistlers, although they often didn't know what they were hearing.

Many beginning enthusiasts use an audio amplifier with a very long wire antenna to hear natural radio signals. In rural environments such an arrangement may work well, but in most cases, input filtering is required to reduce 60 Hz power line hum to a tolerable level. Even with filtering, it is advisable to stay as far away as possible from power lines.

Plans to build natural radio receivers from scratch have appeared in numerous hobby publications over the years. Some notable examples are given in *Chapter 8*. Some commercial receivers are available, and these are also discussed.

TUNING IN EARTHQUAKES

Animal behavior can help predict earthquakes. Animals can detect signals from a quake in progress and tend to run away from the epicenter. The reason may be electromagnetic waves of very low frequency, to which animals are sensitive. Experiments in which specially designed radio receivers detected vibrations in the 10 to 1,500 Hz range (and, surprisingly, around 81 kHz) just prior to major quakes, tending to reinforce this theory.

How would an earthquake give off electromagnetic radiation? The prevailing theory holds that rock formations at fault lines generate electromagnetic waves when they are stressed. It is also believed that the Earth's magnetic field is distorted at these fault lines when a quake is due, giving rise to sudden ionospheric disturbances (SID's) that change the ionization density of atmospheric layers, in turn affecting radio wave propagation.

In addition to signals in the 10 to 1,500 Hz range and at 81 kHz, strange emissions have been noted preceding quakes at other portions of the spectrum including the vicinity of 7 MHz, 9 MHz, 14 MHz and on into the VHF region. Such emissions have been variously described as rain-like noise, interference, abnormal emissions and just plain noise.

Sitting next to a radio receiver tuned to 10 to 1,500 Hz or other frequencies waiting for something to happen won't be productive of course – quakes are too sporadic and infrequent. Perhaps an ink recording device could be used in conjunction with receiving equipment. Such a project is left to the imagination of the reader.

There is another method of quake detection that is interesting and worthy of experimentation. The September/October, 1983 issue of *Monitoring Times*, a monthly magazine devoted to all forms of communications monitoring, told of various experiments being conducted in China, one of which monitored electrical ground currents constantly circulate within the Earth. In this setup, one carbon and one lead electrode were inserted on to two meters into the ground separated by about 20 meters. The two electrodes were wired to a microammeter. Analysis of changes in the currents apparently showed promise in predicting impending quakes.

BRAIN WAVES

The brain generates waves of 1 to 22 Hz. Different frequencies are associated with different bodily functions such as sleep, relaxation, thinking etc. These waves are detectable by sophisticated medical equipment like the electroencephalograph (EEG) and even the simple, inexpensive brain wave monitors found in hobby stores and catalogues.

The fact that these waves are produced raises interesting, and in some cases frightening, possibilities. Perhaps sleep or relaxation, for example, could be induced by generating electromagnetic waves of the appropriate frequency into the brain. Or mood swings might be altered to help those with depression.

The other side of the coin however, it that minds might be "controlled," or even destroyed. Frightening as the prospect may be, experiments have been conducted along these lines, many for military purposes.

This raises the larger question – whether radio transmissions of very low frequency now in use or in the planning stage might affect large number of people over a wide area. For example, could Project ELF discussed earlier have had adverse effects on the human brain, or that of animals? Many feel the answer is "yes". But there is no conclusive proof as of this writing.

Chapter 8
Unusual Uses for Longwave

COMMUNICATIONS UNDERGROUND

Just as the lower frequencies are used to communicate with submerged submarines, they can pass signals through the Earth to explore caves.

A system known as MIDAC (Magnetic Induction Direction Finding and Communications) employs a beacon transmitter operating at 1,200 Hz using an induction coil up to five feet across as an antenna. The transmitter and antenna are carried underground by cavers as they explore the unknown. Receiving apparatus on the surface is tuned the same frequency tracks the explorers as they move below.

When the underground team wishes to have exact depth and locations plotted it stops the beacon transmission and signals to the surface in CW. By taking direction-finding readings from various points on the surface, the position of the cavers below can be determined with high accuracy, making it possible to plot precise cove configurations. Such activity is possible at depths of up to 1,200 feet.

If ongoing communication between the surface and subsurface crews is required, transceivers are used and Morse code is employed. Voice communications have been tried, but the size of the induction coil needed and the power levels required make such operation normally impossible.

In 2006, Los Alamos National Laboratory announced development of a commercially viable technology enabling through-the-Earth communications for mine-rescue applications. The system uses very low frequency electromagnetic radiation and digital audio compression technology to carry voice and text data. The VLF signals can also be used to track and locate the system's users if they are unable to respond. Such a system could have been valuable in underground instances such as the London subway bombing or the Sago mining tragedy.

HAARP

Not all longwave work is below 500 kHz. Since the early 1990s, the Pentagon has conducted an ambitious research program in Gakona, Alaska, called HAARP (High Frequency Active Auroral Research Program). HAARP is jointly managed by the Air Force and the Navy. Among other objectives, the program seeks to generate ELF (Extreme Low Frequency) energy by exciting small parts of the ionosphere with high frequency (2.8-10 MHz) radio energy. It is hoped that the resulting ELF signals (below 5 kHz) can then be modulated and used to provide improved military communication with submarines.

The power levels used at the HAARP facility are impressive. The site's 30 separate transmitters can deliver 3.6 million watts to the antenna array. The antenna system consists of 180 towers and provides enough gain to produce an effective radiated power (ERP) well above 1 *billion* watts. This makes HAARP one of the most powerful radio transmitters in the world.

In addition to communication uses, scientists hope to use HAARP for environmental studies. Recreating the natural effects that occur in the ionosphere due to the sun's stimulation may yield a better understanding of global warming and ozone depletion.

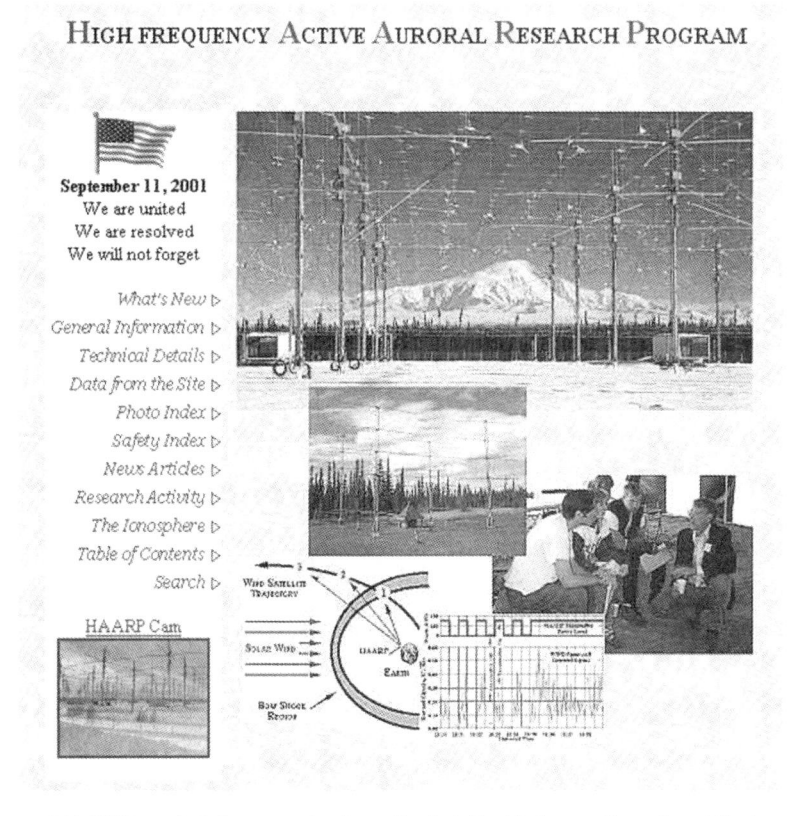

HAARP maintains a web site with detailed information about their operations (www.haarp.alaska.edu).

INDIRECT SOLAR FLARE DETECTION

Most of us picture solar flares as fiery protrusions from the surface of the sun observed through special telescopes. Visual observation is indeed the most common way to "see" solar flares, but another method of indirect detection is just as fascinating.

Solar flares profoundly affect on the lowest region of our ionosphere, the D layer. You will remember that radio waves of the lowest frequencies propagate in duct-like fashion between this layer and the earth's surface. Flares increase the ionization of the D layer, enhancing the signal strength of radio transmissions below 30 kHz.

Transmissions from radio stations operating below this frequency (and there are numerous powerful ones around the world) can indicate solar flares when their signal strength increases. Observation is carried out with strip-chart recorders that record the strength of a selected transmission. A flare produces a bump on the chart, indicating an SES (sudden enhancement of signal).

The activity described is a significant part of the work carried out by a group of researchers in Cambridge, Massachusetts. If indirect solar flare monitoring interests you, send an SASE to the American Association of Variable Star Observers (AAVSO), 25 Birch Street, Cambridge, MA 02138-1205, and request information on AAVSO's Sudden Ionospheric Disturbance program. The group also maintains a web site at www.aavso.org.

AVALANCHE BEACONS

One of longwave's strengths is its ability to penetrate obstructions that would otherwise block radio communications or severely reduce signal strength. The military's use of both VLF and ELF for submarine communication provides two such examples. Another relatively new application for longwaves are personal avalanche beacons. These are small units carried by skiers and hikers that are capable of operation on internationally-standardized frequencies of 2275 Hz (2.275 kHz) or 457 kHz.

Someone trapped in an avalanche, can activate the beacon, which sends a distress signal to properly equipped searchers. Of the two frequencies used, 457 kHz provides somewhat better transmission through deep snow and is the frequency preferred by most users. Research has shown that wearing a beacon increases the odds of avalanche survival by 25 to 35 percent. These numbers may not seem encouraging, but every bit counts in a crisis.

Chapter 9
Receivers and Converters for Longwave

Until the 1980s, hobbyist receiving equipment for frequencies below the AM radio band was practically nonexistent. Now, though, more and more receivers are appearing with coverage down to 100 or 150 kHz. It must be noted that even though a receiver may be able to tune longwave frequencies, it does not always follow that they are sensitive in this region.

Currently available receivers for longwave run the gamut of price, features and performance. At the high end are tabletop receivers, most of which come with a long list of features such as selectable bandwidth (important for separating stations close in frequency), excellent sensitivity, programmable memories, notch filtering and high quality audio. These features can be important to the serious utility monitor, but are not essential for getting started in longwave monitoring.

The R75 is a popular offering from Icom. Unlike most other receivers in its class, it goes down to 30 kHz. Available for about $600 new, $400 used.

The Drake R8B is an example of a full-featured tabletop receiver offering longwave coverage down to 100 kHz. It is well-suited for beacon hunters wishing to focus on the 190-535 kHz band. Earlier R8 and R8A models are also highly regarded for longwave work. All three versions are out of production, but may be obtained on the used market ($500-$900 used, depending on version).

Yaesu's FRG-100 tunes down to 50 kHz and offers a host of features normally found only on more expensive models. Available for about $600 new, $400 used.

Listening to Longwave

Portable receivers offer an excellent and less costly way to get started in longwave listening. They can also serve as a standby receiver for more experienced listeners. Indeed, for some aspects of the hobby, such as listening while traveling or at DX camps, a portable receiver is often the tool of choice.

Today's portables offer features that rival the performance of some tabletop models, at a fraction of the cost. Most models provide longwave coverage down to at least 150 kHz, which is perfect for beacon hunting. For serious DXing, you'll want your receiver to have CW/SSB capability. Besides allowing you to copy CW and data signals, this will help when trying to hear weak signals, or when sifting through closely spaced stations.

Most portables have a built-in ferrite antenna that can provide surprisingly good results on longwave. These antennas exhibit sharp nulls off the ends, so they can be used to peak a desired signal or eliminate an interfering one. Of course, you're not restricted to using the built-in antenna. Most sets include an external antenna jack, or you can use a magnetically-coupled adapter to connect a long wire or other external antenna to the radio.

Sony's ICF-2010 is considered the "King of Portables". It has a proven record for utility reception on both shortwave and longwave frequencies. It tunes down to 150 kHz. Although it was recently discontinued by Sony, it is often available on the used market at $250-$400.

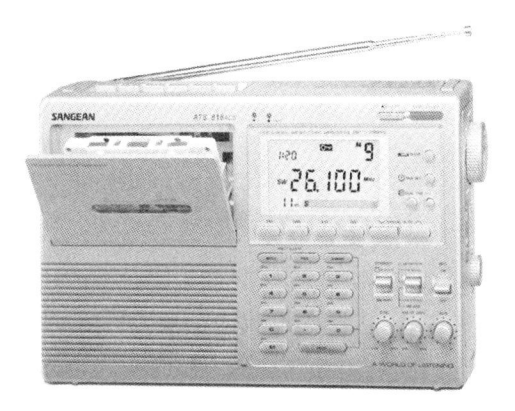

The Sangean ATS-818CS in the $200 price class, is a general purpose portable used by many longwave DXers. It includes a built-in cassette recorder — useful for recording signals off the air. (The ATS-818 is identical, but omits the cassette feature.)

The RF Systems AA-2 magnetic antenna coupler makes it possible to connect an external longwave or medium wave antenna to any portable radio — even if the set lacks an external antenna jack.

Listening to Longwave

Surplus and Vintage Receivers

Yet another way of getting on the longwave band is with an older receiver originally intended for military or professional use. Such sets generally used vacuum tubes and are rather large and heavy. But their design and construction was of the highest quality, with a generous margin of "overkill" in their specifications.

Still, many of these sets date from World War II or not long after, and often show signs of age. Prospective users should be aware of possible dried-out or cracked insulation on wires, corrosion on or under the chassis, worn-out controls and other components. Tubes and capacitors in these older rigs seem to be especially prone to the ravages of time and may have to be replaced to solve problems of hum and audio distortion.

It may be difficult to locate exact replacement parts for these sets, but suitable substitutes can be found in most cases. There are also dealers who specialize in supplying tubes and other components for vintage sets. Antique Electronic Supply (6221 S. Maple Ave., Tempe, AZ 85283) and Radio Daze (7620 Omnitech Place, Victor, NY 14564) are noted suppliers.

The National RBL-5 is a highly respected unit covering 15 to 600 kHz. This model was produced under contract for the U.S. Navy during World War II. It is sometimes seen on the used market at reasonable prices.

Restoring classic sets can be part of the fun of the hobby. Indeed, a properly working military receiver from 55 years ago can rival the longwave performance of even today's best receivers. The lack of digital circuitry in these sets means that they generate less internal noise and thus tend to be "quieter" than modern receivers and can be used for longer periods without listener fatigue occurring.

If using a vintage rig sounds appealing, go for it. Just be aware of the pitfalls mentioned above, and be prepared to expend some effort getting your receiver into top notch shape. Hamfests, swap meets and surplus dealers are the best places to look for older gear. If you're lucky enough to get a manual with your purchase, consider it a bonus. Otherwise, there are a number of aftermarket firms that carry manuals for older sets. Check the Reference Section for a list of suggested sources.

The Collins R-389/URR is considered by many to be the finest vintage LW receiver available. It covers 15 kHz to 1500 kHz. These models, produced from 1951 to 1955, have become rather scarce.

The Hammarlund "Super Pro" SP-600-VLF-31/38 is a quality longwave receiver covering 10 to 540 kHz. Its long production run speaks for itself—1951 to 1972.

Longwave Converters

VLF converters offer another approach to getting on the longwave band. If you already have an older shortwave receiver, or one that does not tune low enough for your interests, a converter may be an excellent choice. Most converters provide reception of longwave signals down to about 5 kHz.

Converters work by "moving" a portion of the longwave band to a range of frequencies commonly covered by a shortwave receiver. They are installed in-line between the antenna and receiver and require no modifications to the receiver whatsoever.

Common output frequency ranges are 3.0 to 3.5, 3.5 to 4 MHz and 28 to 28.5 MHz. This means that a converter can also be used with most ham band transceivers, provided steps are taken to prevent transmitting into the converter. (This would cause instant damage to the converter circuitry.)

You don't have to purchase a converter – you can build your own. These circuits are usually uncomplicated, requiring only a handful of "junk box" parts and soldering skill. Numerous construction articles for LF converters have appeared in hobby publications over the years. A notable example was in the June 1997 issue of *Monitoring Times* magazine. Converter plans are also offered by Panaxis Productions. See the Reference Section for contact information.

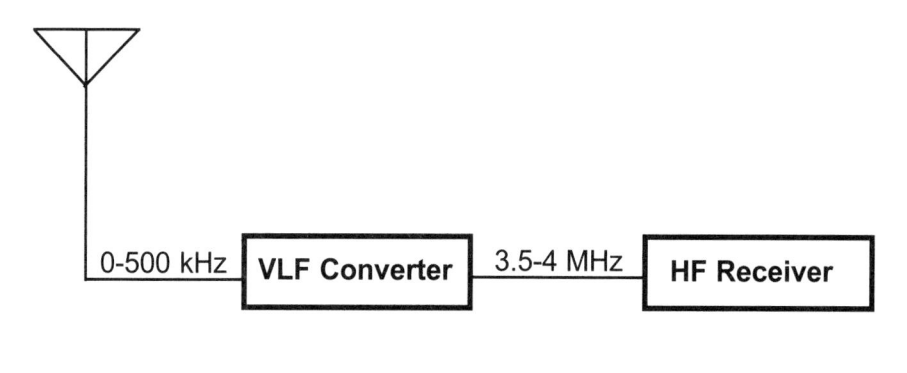

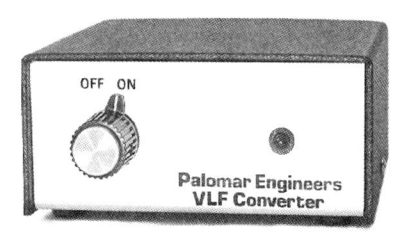

```
       ▽
       |
       |
 0-500 kHz  [ VLF Converter ]  3.5-4 MHz  [ HF Receiver ]
```

The Palomar VLF Converter (about $100) has been a favorite for many years.

Listening to Longwave

Natural Radio Receiving Equipment

The study of radio signals produced by nature has been a fascinating part of scientific research since the mid 1950s. In recent years, however, natural radio has enjoyed popularity in hobby circles as well. It represents an area where amateurs can make valuable contributions to research using simple receiving equipment. Several types of natural radio signals and their causes were covered earlier, in Chapter 5. We'll now explore some of the equipment options for the experimenter interested in this part of the band.

Some experimenters homebrew their natural radio receiving equipment. The circuits are usually uncomplicated and can be made with easily obtained parts. Construction articles have appeared in recent publications, for example:

> "Whistler VLF Receiver," 1990 *Popular Electronics Hobbyist Handbook*

> "Wideband VLF Receiver," *Radio Science Observing* by Joseph J. Carr

> "IC Whistler Receiver," *The Lowdown,* September 1994

> "Ferrite Rod Sferics Receiver," *The Lowdown,* September 1998

> "Bare Bones Basic (BBB-4) Receiver," *Monitoring Times* March and April 2006 issues

As of press time, plans for two types of whistler receivers exist on the Longwave Club of America's web site at: www.lwca.org. Check the library files.

If homebrewing does not interest you, ready-made equipment is available for natural radio listening. Two prominent firms involved in this market are LF Engineering Co. and Kiwa Electronics. Both firms offer small, portable units capable of hearing signals with minimal fuss. These models also have a provision for connecting a tape recorder to preserve your most interesting catches.

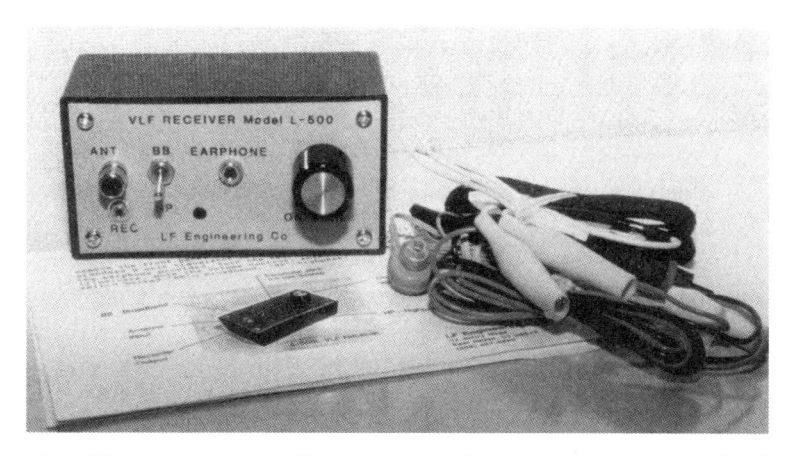

The LF Engineering L-500 comes ready for operation on natural radio frequencies. It includes a belt loop carrying strap, earphone, and tree-tapping probes.

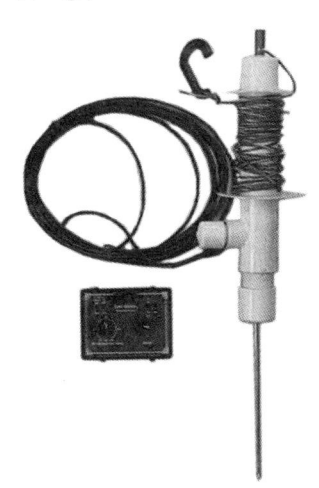

The Kiwa Earth Monitor features adjustable filtering, and includes a spool of antenna wire and a built-in grounding probe. As with all natural radio receivers, it can only be effectively used far away from power lines. This model was no longer in production at press-time.

Listening to Longwave

Chapter 10
Longwave Lowfer Transmitters

For those hardy souls willing to brave the challenges of the 1750 meter Lowfer band (160-190 kHz), equipment availability is just another hurdle to deal with. From time to time, various firms have made transmitters for this band, but the selection has always been slim. This situation may improve if the pending proposal for a ham band on 1750 meters (or a band at 136 kHz) is approved by the FCC.

In the mid 1990s there was at least one Lowfer transceiver available in kit form—the EXP-1750. It was offered by Lowfer experimenter David Curry, WD4PLI, of Burbank, California. This unit was capable of running CW or single sideband (SSB) and covered the entire 160-190 kHz band. A small group of experimenters in California still use this transceiver to communicate on a regularly scheduled network.

Although no longer available commercially, a construction article for this transceiver appeared in the April 1994 issue of *QST Magazine* (page 26). Reprints are available from the ARRL Headquarters of a small fee. Reprint information can be found at: *www.arrl.org*.

The EXP-1750 was a full-featured CW/SSB transceiver previously available in kit form. (photo by Dave Curry)

Another construction article for a basic (CW) Lowfer transmitter appeared in the July, and August 1998 issues of *Monitoring Times* magazine. Designed around a handful of parts, mostly from Radio Shack, it would provide a practical way to get started on the Lowfer band or at 136 kHz.

The EXP-1750, shown in its optional enclosure, is no longer available commercially. See previous page.

Fortunately, Lowfer experimenters have been quite willing to share their transmitter designs, and many of these circuits have appeared in hobby publications in recent years. Readers wishing to construct a Lowfer transmitter are encouraged to review back issues of *The Lowdown*, which contain a wealth of information on transmitters and accessories. A circuit for a basic transmitter was also presented in the June and July 1998 issues of *Monitoring Times*. A must-have reference for those engaged in building their own gear is the *LF Experimenters Source Book* mentioned earlier.

Chapter 11
Longwave Antennas

Antennas for receiving longwaves are as important as the receiver itself. Major categories include verticals, long wires, loops and active whips. No one type is necessarily better than the other. Different conditions will show a certain antenna performs best one time, while another does so at a different time. Two primary problems with antennas for the VLF spectrum are that the wavelengths involved are so long that erecting a dipole or other resonant antenna is all but impossible, and there is the ever-present problem of noise.

Verticals are a popular choice because they provide good groundwave reception. In addition, many Lowfers make use of these in the form of a 15 meter vertical for transmitting in the 1750 meter band. A good radial ground system must be used with this type of antenna. The major drawback to verticals is susceptibility to noise pickup.

Long wires ranging in length from 50 feet on up are also a popular choice, but these can be noisy too.

Loops are antennas shaped in the form of circles or squares. They can be a single or multiple-turn coil of wire and vary in diameter from less than a foot to five feet. Their biggest advantage is the fact that they exhibit a figure-8 reception pattern, so rotating them can null out interfering noise.

Active whips have become very popular in recent years. These devices consist of a short whip approximately 1 meter long, connected directly to a preamplifier. A cable from the preamp runs to a coupling device that supplies operating power and couples the signal to the receiver.

An in-depth discussion of various longwave antennas was published in a series of articles by R. W. Burhans in the February through June, 1983, issues of *Radio-Electronics Magazine*. The articles are excellent, but reprints may be difficult to find. A more recent discussion of longwave antennas can be found in the *LF Experimenter's Source Book* and *Joe Carr's Receiving Antenna Handbook*. See the Reference Section for contact information.

The myriad of antenna types should not stop you from getting started. A 50 foot length of wire strung out the window as high as possible will serve to start you on some great adventures until you design that perfect wire in the sky.

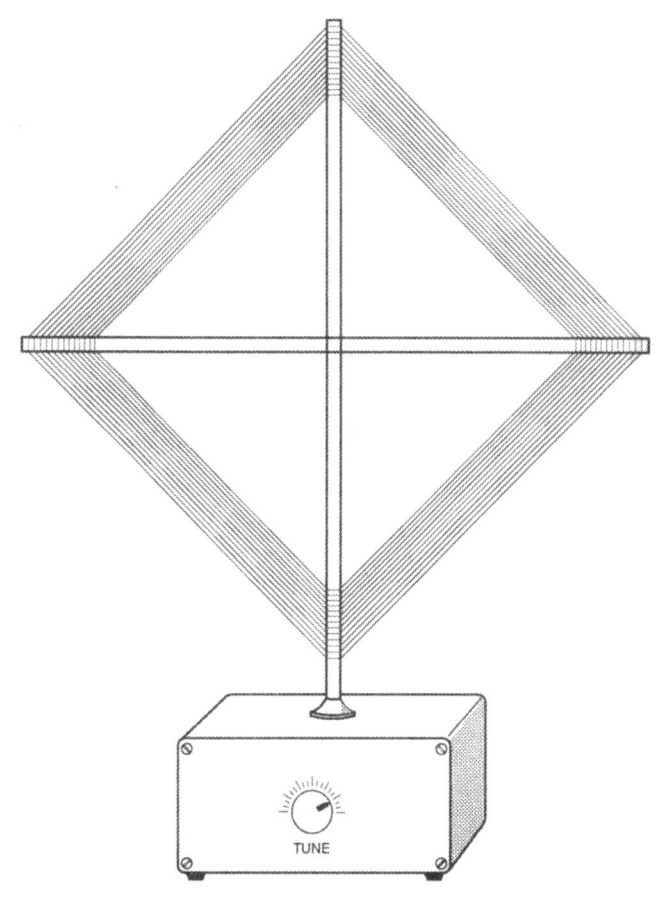

A generic wire loop antenna for longwave reception.

A special type of loop antenna is the ferrite bar. Although these are often built into portable receivers, reception can be greatly improved through the use of a tuned, external ferrite loop. As with any other loop antenna, a ferrite bar can be rotated to null out interference or peak a desired signal. It can also be used for direction finding.

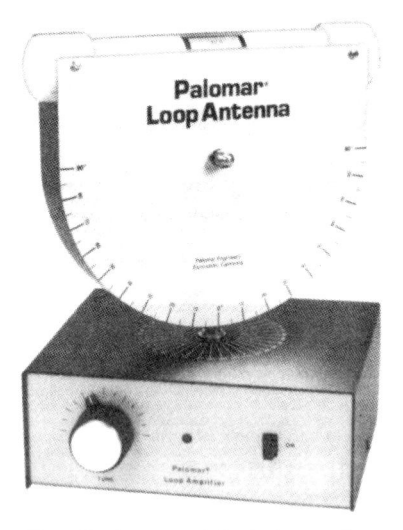

The Palomar LA-1 loop base shown with loop element installed.

For many years, Palomar Engineers has manufactured the popular LA-1 amplified ferrite antenna. A large selection of plug-in loops are offered with the LA-1 base to cover from 10 to 15,000 kHz. The LF Loop Element covers 150 to 550 kHz and will be of most interest to longwave listeners.

Active antennas are omnidirectional and are a good choice for a general-purpose receiving antenna. They are known for their low-noise reception, and take up much less space than a wire antenna of comparable performance. The best active antennas include filtering to attenuate signals above 550 kHz. This could be important if you live near an AM broadcast station.

The LF Engineering L-400B is an active antenna specifically designed for the longwaves (10-530 kHz).

Active antennas require only a light-duty support for installation such as a small mast, vent pipe or gutter. They're easy to move around, so try several locations for the best reception and lowest noise before choosing a permanent mounting spot. Sometimes a move of just 20-30 feet can make a big difference in noise pickup.

The Par BCST-LPF low pass filter is designed to help long wave DXers cope with interference from stations above 540 kHz, especially nearby AM broadcast stations. Strong AM stations may overload the front-end of your receiver. This may cause an AM (medium wave) station to appear on the long wave band. The Par BCST-LPF is a 5 pole elliptic low-pass filter designed to reduce and often eliminate AM broadcast problems in the 0 to 500 kHz range. The housing has coaxial SO-239 connectors for input and output (to accept standard PL259 plugs). A switch is featured to bypass the filter. The filter's bandpass is 0 to 500 kHz. The stop band is 540 kHz to 300 MHz and the minimum attenuation for 500-1600 kHz is 32 dB.

The Par BCST-LPF filter (about $60) reduces interference on the longwave band from nearby AM broadcast stations.

While any hunk of wire strung to a tree can be pressed into service as a receiving antenna, *transmitting* is a different story. With Lowfers strictly restricted in power and antenna length, a transmitting antenna must be as efficient as possible if a station is to be heard more than a few miles away.

A time-proven design used by many Lowfers is the top hat vertical. This antenna consists of a guyed tube 25 to 40 feet in height, with a capacity "hat" attached at the top. A "push-up" telescoping mast is often used for the vertical portion of the antenna. These masts are commonly available at electronics stores and home centers that carry TV antennas and mounting hardware.

Tophat verticals are used by many successful Lowfers and at some newer aviation beacon sites.

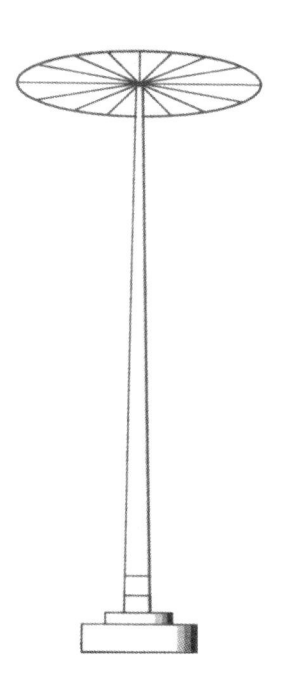

The transmitting antenna must be well insulated from the ground to avoid losses and arcing that can occur from the high RF voltages involved. In addition, a loading coil is required for tuning the antenna to resonance. The coil should be placed in a weather-tight enclosure to avoid changes in inductance from rain or moisture. Many operators use a plastic tool box or garbage pail at the base of the antenna to house the loading coil.

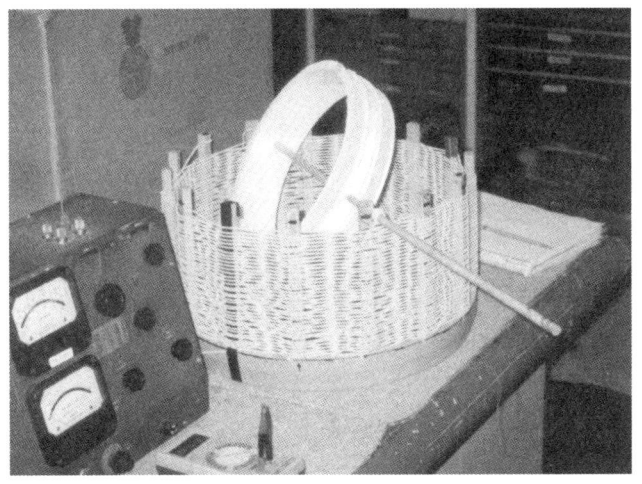

An example of a Lowfer loading coil under construction. (Photo by Bill Bowers)

The ground radials are another crucial part of the antenna system. These should be as long as possible (at least as long as the antenna is high) and should extend out from a central grounded hub like the spokes of a wheel. There is no rule for how many radials should be used, but generally six is the minimum, and the more and linger is better. Some operators tie radials to other grounded objects, such as well casings and chain link fences, to further improve performance.

Those interested in building a transmitting antenna may wish to consult the October 1999 issue of *The Lowdown,* which carried complete plans. In addition, Panaxis Productions, mentioned earlier, also carries plans for building an antenna.

Chapter 12
Longwave Listening Tips

Monitoring the basement band can be challenging, especially for the newcomer. A few tips can help make the road a little smoother. This chapter is a potpourri of miscellaneous information designed to improve your listening success. The more you learn about the VLF spectrum the more you will enjoy it. One of the best ways to learn, of course, is to read.

The *Longwave Club of America* was mentioned as a must for those interested in Lowfer activity. Actually, it's a must for anyone interested in the frequencies below 500 kHz, for this portion of the spectrum is the very essence of what the club represents. Its monthly newsletter, *The Lowdown*, contains authoritative articles on every subject concerning the longwaves. The LWCA also maintains an informative website: **www.lwca.org**

The Lowdown is the monthly publication of the Longwave Club of America.

THE
LOWDOWN
A PUBLICATION OF THE LONGWAVE CLUB OF AMERICA

PAGE

1 LOGGINGS – DX DOWNSTAIRS
   Bob Montgomery

7 THE LF NOTEBOOK
12 THE 1750 METER BAND
14 THE TOP END
   John H. Davis

15 LF AND VLF CONVERTER UPDATE
   Harry A. Weber

16 NATURAL RADIO
   Mark Karney

20 VLF PRESELECTOR
   Harry A. Weber

22 QSL
   Richard D. Palmer

SEPTEMBER   2005

The *LF Experimenters Handbook,* by Peter Dodd G3LDO contains detailed information and schematics for constructing converters, preamps, filters, wavetraps, antennas, transmitters and more for the basement band. Although the emphasis is on the European 136 kHz amateur band, most of the information is adaptable to other parts of the longwave spectrum. Another book that focuses on this subject is *LF Today - A Guide to Success on 136 kHz* by Mike Dennison G3XCV. Both of these books, published in England by the Radio Society of Great Britain, are available via the A.R.R.L.

Popular Communications and *Monitoring Times* are two excellent monthly periodicals that cover every conceivable aspect of communications "from DC to daylight," as the saying goes. Frequent articles appear in both magazines covering the very low frequencies. Since 1988, *Monitoring Times* has featured a monthly column on longwave called *Below 500 kHz*.

Purchasing or joining or subscribing to these books, periodicals and clubs would guarantee you wouldn't miss much information on longwaves.

Concerning Antennas

Get them as high and in the clear as possible. This may sound obvious, but it really applies in this case. The author has found his 2-meter, 20-element collinear, mounted at 50 feet and in the clear, provides better (more noise-free) reception that his 67 foot dipole which is only 20 feet above ground. No suggestion is being made that VHF antennas be used for VLF reception; a point is simply being emphasized.

Antenna Tuners

Use an antenna tuner for long wires and dipoles. This can dramatically improve reception by providing a better impedance match and reducing broadcast band interference that can plague the lower frequencies. The lower you go in frequency, the more such a device will help. It is possible to realize improvements in signal strength of 20 dB or more as the very lowest frequencies are tuned.

The MFJ-956 Preselector/Tuner (about $60) boosts desired signals while rejecting images, intermod and other phantom signals. It operates from 150 kHz to 30 MHz.

Audio Filters

Due to limited power and restricted antenna lengths, most Lowfer signals (except those close by) are very weak. It requires some real digging to pull these experimenters out of the noise. One of the best ways to is to use the narrowest bandwidth receiving equipment you can get your hands on. This means receivers with narrow I.F.'s and the use of CW audio filters. The latter piece of equipment can be homebrewed if you like, but it is also available from many manufacturers. It is especially popular among CW enthusiasts.

The MFJ-752C (about $110), is an example of an analog-based audio filter useful for sorting out interfering signals.

Digital Signal Processing (DSP) filters, such as the Timewave DSP-599zx (about $440), are powerful tools for eliminating interference and enhancing desired signals

Tips for Beacon Hunters

With a few exceptions, beacons use Morse code characters for their identifiers. If you don't know Morse code, don't worry. The idents are sent so slowly and repeated so many times you can simply jot down the dots and dashes as you hear them. Then, look up the letters or numbers they represent in the Morse code list on page 9 of this book.

Be aware that beacons malfunction some of the time. The keying devices used to repeat their I.D.'s over and over again can fail and result in missing or poorly sent characters. A dash may end up being too short, a dot too long. Or a character might be dropped completely. Sometimes this is easily noticed when you hear an identifier that makes no sense whatsoever; that is, characters being transmitted that do not represent any legitimate character in the Morse code. At other times the failure may be more difficult to detect. If the "E" were dropped from "MME," it could go unnoticed. You might think you've discovered a new beacon instead of a faulty one.

In rare cases, a beacon may be heard transmitting its ID with "negative keying." Usually, you won't have any trouble recognizing negative keying. It is a peculiar transmitter defect where the ID is sent out as an "upside down" image of the true ID. Wherever there's a space in the true ID, there will be a tone in negative keying and vice-versa. Because of this inversion, the ID you hear will likely be a random series of dots and dashes that make no sense at all. For example, the negative keying of beacon GLY would be a 5-letter ID; "ERUIE." Most IDs have between 1 and 3 characters, so this is a sure tip-off that something is wrong.

Should you encounter a negative keyed signal, here's an easy way to convert it back into Morse: Plot the negative keyed sounds on a sheet of graph paper so that single blocks represent dots, and three blocks represent dashes. Assign one block to the space between individual dots and dashes and three blocks to the spaces between complete letters. Next, directly below your plot, fill in the blocks where there is white space above and read out the true ID.

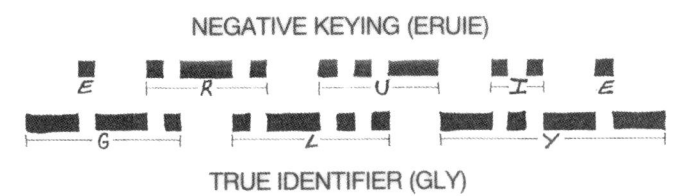

NEGATIVE KEYING (ERUIE)

TRUE IDENTIFIER (GLY)

Negative keying mysteries can be solved by plotting the ID sounds on graph paper. Here's an actual example used to solve the negative keying of beacon GLY at 388 kHz in Clinton, Missouri.

Another clue to negative keying is that there will be a dash after the ID on U.S. beacons. Normally, U.S. beacons do not have a dash. Conversely, Canadian beacons, which usually *do* have a dash, will have a long space instead.

Longwave DXer's Checklist

It is important to keep in mind the ever-changing state of the radio spectrum. As new beacons appear, others change frequency, and some are put out of service. A number of publications can help determine the identity and location of those you hear. One of these is the "Beacon Finder" which lists hundreds of beacons and other used of the longwave band. Another good source is Sectional Aeronautical Charts which are maps showing callsigns, locations and frequencies of aircraft beacons. These are listed in the Reference Section. A number of online resources exist as well.

Whether you're interested in chasing distant beacons, LF broadcasters or weak-signal Lowfers, these time-proven hints will give you the best chance of success. The items below can be used as a checklist to ensure you hear the most signals possible.

1. Use a low-noise receiving antenna such as a loop or active antenna.
2. Choose a narrow bandwidth setting (500 Hz or less) to reduce interference from adjacent signals.
3. Use an outboard audio filter (analog or DSP) to eliminate unwanted tones and interference.
4. Use the CW or SSB mode on your receiver (BFO on).
5. If possible, shut off sources of household interference (fluorescent lights, dimmer switches, electric motors, TV sets, computers, etc.)
6. Tune slowly to avoid missing signals. Stations are normally assigned to 1 kHz channel spacings.
7. Use a good set of communications headphones to block out household noise.

Choose your listening times carefully. While there are exceptions, the best times tend to be at night, and during the cooler months.

Logging Your Catches

A logsheet will chart your DXing progress. It gives you something to show for your efforts, and will help you spot changes that occur on the band from year to year. A log need not be anything fancy. You can make up a ruled sheet and make photocopies as needed. If you want to go first class, a logsheet could be kept on a computer. Whatever the format of a log, there are some special entries that may be of interest to the longwave DXer. Of course, there's the usual date, time, frequency, ID, signal strength and location of the station heard. Other entries to consider are:

ID Pitch—The two tones you'll hear from most beacons are 400 Hz and 1020 Hz. Traditionally, U.S. beacons use 1020 Hz, while Canadian beacons use 400 Hz. There are some exceptions to the rule, where just the opposite is true...all the more reason to have it down on paper!

Distance—The distance in miles (or kilometers) from your listening post to the beacon site is very useful information. One technique for measuring distance is to post a map in your shack marking your location with a thumbtack. Attach a thin, movable strip of paper marked off in miles/km to the thumbtack. The strip can be rotated in any direction to measure distances.

Beacon Power—To put a logging in perspective, it's important to know the transmitting power of the station. For instance, hearing a 2000 watt beacon 500 miles away may be fairly routine, but pulling in a 25 watt beacon at that distance would be a good catch by any standards. Some listeners go so far as to calculate a "miles-per-watt" figure for DX catches.

Service—In this space you could have a code letter to signify the type of station you heard (i.e., A=aeronautical, L=Lowfer, B=broadcaster)

IDs per minute—This is the number of *complete* identifications sent by the beacon in one minute. This represents the "fingerprint" of a beacon and it can be helpful to include this information in a verification letter as proof of reception.

Remarks—This is a catch-all area for noting things such as the weather conditions at the time of reception, QSL information, keying errors, etc.

Miscellaneous Tips

Don't be discouraged if your first attempt at longwave listening isn't overly successful. If you begin listening at 12:00 noon on a summer day from an urban location as thunderstorms approach, you won't hear anything but static. Noise is your biggest enemy. Winter always provides better reception than summer, and winter nights allow some of the best listening of all.

Since man-made noise also contributes greatly to the static problem, rural locations are better than urban ones.

Of course, you're not going to move just for better VLF reception (although there are those who have), but some things can be done to improve the situation if you are a city-dweller. Try the use of a good AC line filter; this can help. Some go so far as to operate their VLF gear from batteries, such as a spare car battery, assuming of course the equipment will accept 12 volt DC power. Others go on DXpeditions to remote, quiet, country hilltops to see what signals they can snag from these vantage points. Sound crazy? Not really – it's loads of fun!

Chapter 13
References

In order to eliminate the possibility of outdated pricing information, no costs are listed for items in this section. Write directly to the address shown for particulars on that in which you are interested.

Clubs, Groups and Organizations

The Longwave Club of America is devoted to the promotion of DXing and experimenting on frequencies below 550 kHz and activity on the 1750 meter band. *The Lowdown* is the monthly newsletter of the Longwave Club of America.

> The Longwave Club of America
> 45 Wildflower Road
> Levittown, PA 19057
> www.lwca.org

For information on solar flare detection, contact AAVSO.

> American Association of Variable Star Observers (AAVSO)
> 25 Birch Street,
> Cambridge, MA 02138-1205
> www.avso.org

The *Amateur Radio Research and Development Corp. (AMRAD)* is a group of amateur radio experimenters specializing in telecommunications technology. They have a large emphasis on LF topics and maintain a web page devoted to projects and other happenings on the frequencies below 500 kHz.

> AMRAD
> P.O. Box 6148
> McLean, VA 22106-6148
> www.amrad.org

Books and Other References

LF Experimenters Handbook by Peter Dodd G3LDO contains detailed information and schematics for constructing converters, preamps, filters, wavetraps, antennas, transmitters for long wave. (Cover is shown on page 81). Available in Europe from the R.S.G.B. or in the USA from:

> American Radio Relay League
> 225 Main St.
> Newington, CT 06111
> www.arrl.org

LF Today - A Guide to Success on 136 kHz by Mike Dennison G3XCV is a resource for amateurs operating on the longwave band. Available in Europe from the R.S.G.B. or in the USA from:

> American Radio Relay League
> 225 Main St.
> Newington, CT 06111
> www.arrl.org

The Beacon Finder II by Kevin Carey lists hundreds of radio beacons and other longwave utility stations between 0 and 2,000 kHz. Available from:

> Kevin Carey
> P.O. Box 56
> West Bloomfield, NY 14585

Sectional Aeronautical Charts feature maps that cover various sections of the country. They list aircraft beacon callsigns, locations and frequencies. Available from your local airport or:

> Distribution Division (C44)
> National Ocean Survey
> Riverdale, MD 20737

VLF Radio Engineering by Arthur D. Watt is published by Pergamon Press. This is a technical book, some 700 pages in length, often found in engineering school libraries. There is a good chance it can be obtained from your local public library through an inter-library loan. Its objective is to provide "detailed coverage of the fields involved in VLF radio engineering, a compendium of basic antenna, propagation and system engineering information, and a guide for applying the information in the solution of practical problems." Only those wishing to pursue the subject in depth will be interested in this text, but if such is your pleasure this book will prove to be invaluable.

The World Radio TV Handbook. This venerable annual publication provides frequencies, schedules and other information for all longwave, medium wave and shortwave broadcast stations. This book is carried by radio dealers and major book stores worldwide.

The Low & Medium Frequency Radio Scrapbook by Ken Cornell, W2IMB (deceased) saw ten editions printed over twenty-five years, starting in 1972. These guide offered practical information on transmitters, receiving gear, antennas, and other equipment for the LW enthusiasts. Packed with schematics, coil winding data, regulatory information, and other valuable material. Although out of print since 1977, used copies can sometimes be found at swapmeets, hamfests and online auctions.

Periodicals

MT is a monthly magazine devoted to all facets of radio monitoring.

Monitoring Times
7540 Hwy. 64 W.
Brasstown, NC 28902
www.monitoringtimes.com

PopComm magazine is devoted to all facets of radio communications.

Popular Communications
25 Newbridge Road
Hicksville, NY 11801-2953
www.popular-communications.com

Manufacturers & Distributors of Related Products

Low Frequency Engineering Company
17 Jeffrey Road
East Haven, CT 06512
860 526-4759   www.lfengineering.com

MFJ Enterprises
Box 494
Mississippi State, MS 39762
800 647-1800   www.mfjenterprises.com

Antique Electronic Supply
6221 S. Maple Ave.
Tempe, AZ 85203
480 820-5411   www.tubesandmore.com

Radio Daze, LLC
7620 Omnitech Place
Victor, NY 14564
585 742-2020   www.radiodaze.com

W7FG Vintage Manuals
119 E. George Street
Batesville, IN 47006
812 932-3417   www.W7FG.com

Boat Anchor Manual Archive (BAMA)
An online resource for out-of-print manuals and schematics.
http://bama.sbc.edu/

The Manual Man
27 Walling Street
Sayreville, NJ 08872
www.manualman.com

Palomar Engineers
Box 462222
Escondido, CA 92046
760 747-3343   www.palomar-engineers.com

Panaxis Productions
P.O. Box 130
Paradise, CA 95967-0130
530 873-9100   www.panaxis.com

Par Electronics
P.O. Box 645
Cullowhee, NC 28726-0645
828 743-1338   www.parelectronics.com

Autek Research
PO Box 7556
Wesley Chapel, FL 33544
813 994-2199   www.autekresearch.com

Ramsey Electronics
590 Fishers Station Dr.
Victor, NY 14564
800 446-2295   www.ramseyelectronics.com

Grove Enterprises
7540 Highway 64 West
Brasstown, NC 28902
828 837-9200   www.grove-ent.com

Universal Radio
6830 Americana Parkway
Reynoldsburg, OH 43068
614 866-4267   www.universal-radio.com

Useful Websites

Keep in mind that web addresses change frequently, and it is possible that some of the sites listed here may have ceased operation entirely since this book went to press. If you encounter a nonworking link, try performing a search that includes key words for the site you are looking for.

http://hometown.aol.com/rkdx/
Many LF loggings, news of DX stations heard, links to other LW and MW sites.

http://www.lwca.org/
Longwave Club of America (LWCA). This is probably the biggest LW site on the net. Widely varied information on all facets of Longwave operation. Includes a message board. Also includes an online version of "The Art of NDB DXing," a series on maximizing weak signal reception of beacons.

http://www.datasync.com/~rocker/longwave.htm
Heavy emphasis on loggings, updates on unidentified stations.

http://www.jazzkeyboard.com/jill/radio/longwave.html
Addresses for many beacons, QSL information, Beacon listings and extensive links to other LW resources.

http://www.qsl.net/sv1xv/lw.htm
Heavy emphasis on beacons in Greece and Europe. Links to numerous other LW sites ranging from natural radio to building your own LF transmitter.

http://homp.planet.nl/~boend177/
Utility DXer's Forum (UDXF). News and information about all types of "Ute" (utility) stations — LF through HF. Includes many related links.

http://worldaerodata.com
World Aeronautical Database. One of the most complete lists of Navaids available on the web, and easy to use.

http://www.dxzone.com/cgi-bin/dir/jump2.cgi?ID=4608
South American NDB Page. Can take a while to load (depending on your connection speed), but worth it. Perhaps the most complete listing of S. American and African NDBs on the Internet.

http://www.spaceweathersounds.com/
Stephen P. McGreevy's Natural Radio Site. THE place on the web for information on all aspects of natural radio. Includes "The VLF Story," a listener's handbook, and sound files of natural radio signals.

http://www.mlecmn.net/~lyle/
An excellent source of information for those interested in the license-free Lowfer band (160-190 kHz). Includes plans for a simple LF transmitter, PC-based identifier and many other projects. Detailed discussion on transmitting antenna design and efficiency. Highly recommended.

http://ca.geocities.com/ve3gop@rogers.com
Alex Wiecek's *Longwave Page* provides wide ranging coverage of the longwave hobby, with an emphasis on Canadian Navaids. Includes many useful links to other online resources.

http://www.beaconworld.org.uk
The *Beaconworld* website is packed with information on DXing for beacons on LF, MF and beyond. Be sure to note the *Beacon Hunter's Handbook* on this site. It is a 200 page PDF file explaining all aspects of beacon DXing.

http://www.classaxe.com/dx
Martin Francis' *DX Radio Pages* is packed with logging tools and databases of NDBs. Be sure to view the NDBRNA list (Non-Directional Beacons Received in North America) on this site. It serves as an extensive list of "Who's hearing what" and can be sorted by many different criteria (time, frequency, location, ID, etc.). A powerful tool.

ACKNOWLEDGMENTS

Allen Renner
Bill Bowers
David Curry
Howard Mortimer
Hugh M. Hawkins
Jill Dybka
Joe Saloka
John H. Cobb, Jr.
Avery Comarow
National Institute of Standards & Technology
Scientific Radio Systems, Inc.
Southern Avionics Co.
Stephen P. McGreevy
The Longwave Club of America
United States Coast Guard

A special thanks is extended to the readers of Monitoring Times *Below 500 kHz* column for their many contributions of loggings, QSLs and other useful material.

The author wishes to especially thank Jacques d'Avignon, VE3VIA, for his support and encouragement during this writing project. Jacques' tireless enthusiasm for the Longwave hobby, his work in organizing longwave DX-peditions, and his willingness to help others in the hobby is greatly appreciated. His efforts serve as an inspiration to all who explore the frequencies below 500 kHz.

ABOUT THE AUTHOR

Kevin Carey resides in the Finger Lakes region of New York with his wife Kirsten and their sons, Bryan and Jordan. He is employed as a technical writer with Microwave Data Systems, a major supplier of wireless telemetry gear used in public safety, utility, and transactional applications.

While he enjoys all facets of the radio hobby, Kevin is particularly interested in the lower reaches of the radio spectrum. This interest led to a freelance writing assignment with *Monitoring Times* magazine, where he has edited the *Below 500 kHz* column since 1991. As a radio amateur (WB2QMY), he also enjoys operating low power "QRP" gear and using vintage equipment on radio's HF bands. In addition to radio, Kevin enjoys outdoor activities such as bird watching, boating, and motorcycling. He has long served as a volunteer fire fighter in his community and is active in other civic organizations. This is Kevin's first book published for the longwave radio hobby.

ABOUT THE PREVIOUS AUTHOR

Pete Carron was schooled in electronics and worked seven years in that field. He switched careers to data processing and has since spent over twenty years in that profession, working today as a self-employed contract programmer and data processing consultant to industry. He has been a licensed amateur radio operator (W3DKV), shortwave listener and communications enthusiast for more than 40 years. In addition, he is author of "Computers - How to Break Into the Field" (Liberty Publishing Company) and "Morse Code - The Essential Language" (A.R.R.L.).